LOCKSMITH AND SECURITY PROFESSIONALS' EXAM STUDY GUIDE

About the Author

Bill Phillips is president of the International Association of Home Safety and Security Professionals. He has worked throughout the United States as an alarm systems installer, safe technician, and locksmith. He is a graduate of the National School of Locksmithing and Alarms (New York City branch), and he currently works as a security consultant and freelance writer whose articles have appeared in *Consumers Digest, Crime Beat, Home Mechanix, Keynotes, The Los Angeles Times*, and many other periodicals. He is the author of the "Lock" article in the World Book Encyclopedia and twelve security-related books, including McGraw-Hill's *The Complete Book of Home, Site, and Office Security*; *The Complete Book of Locks and Locksmithing*, Sixth Edition; *Locksmithing*; *Master Locksmithing*; and *The Complete Book of Electronic Security*.

LOCKSMITH AND SECURITY PROFESSIONALS' EXAM STUDY GUIDE

Bill Phillips

New York   Chicago   San Francisco   Lisbon   London   Madrid
Mexico City   Milan   New Delhi   San Juan   Seoul
Singapore   Sydney   Toronto

The McGraw-Hill Companies

Library of Congress Cataloging-in-Publication Data

Phillips, Bill.
 Locksmith and security professionals exam study guide / Bill Phillips.
   p. cm.
 Includes bibliographical references.
 ISBN 978-0-07-154981-3 (alk. paper)
 1. Locks and keys—Examinations—Study guides.   2. Police, Private—Examinations—Study
Guides.   I. Title.

 TS521.H45   2009
 621.389'28076—dc22

2008037493

McGraw-Hill books are available at special quantity discounts to use as premiums and sales
promotions, or for use in corporate training programs. To contact a representative, please visit
the Contact Us pages at www.mhprofessional.com.

Locksmith and Security Professionals' Exam Study Guide

1 2 3 4 5 6 7 8 9 0   QPD/QPD   0 1 4 3 2 1 0 9 8

ISBN 978-0-07-154981-3
MHID 0-07-154981-1

This book is printed on acid-free paper.

Sponsoring Editor	**Proofreader**
Joy Bramble Oehlkers	Paul Tyler
Acquisitions Coordinator	**Production Supervisor**
Rebecca Behrens	Pamela A. Pelton
Editorial Supervisor	**Composition**
David E. Fogarty	TypeWriting
Project Manager	**Art Director, Cover**
Patricia Wallenburg	Jeff Weeks
Copy Editor	
Marcia Baker	

To my parents, Oscar and Ruby Carr

Contents

Chapter 10 BASIC ELECTRICITY AND ELECTRONICS 71

Chapter 11 EMERGENCY EXIT DEVICES 79

Chapter 12 WIRELESS AND HARDWIRED ALARMS 95

Chapter 13 HOME AUTOMATION 103

Introduction

Locksmith and Security Professionals' Exam Study Guide provides locksmith and security information, and it includes samples of several security-related exams.

If you're a locksmith, alarm system installer, or security officer who wants to get ahead in your current position, you need to be certified—and this book provides you with all the information you need. It includes sample questions from the Certified Protection Professional exam, the Certified Protection Officer exam, the Registered Professional Locksmith exam, the Registered Security Professional exam, and the General Locksmith Certification exam. The answers to the exam questions are in Appendix G.

Appendix D includes an exam you can take to earn Registered Security Professional registration at no charge—a $50 savings.

Even if you don't plan to take a security-related exam, you can still learn a lot from this book. *Locksmith and Security Professionals' Exam Study Guide* contains comprehensive chapters on locks, safes, alarms, closed-circuit television systems, fire safety, securing doors and windows, and safety and security lighting.

Each chapter ends with a quiz—whose answers are in Appendix A—to ensure that you understood the chapter's contents.

If you have any questions or comments about this book, you may contact me via e-mail at locksmithwriter@aol.com, or you may write to me at: Box 2044, Erie, PA 16512.

Acknowledgments

I owe a lot of people thanks for helping me with this book. From conception to completion, *Locksmith and Security Professionals' Exam Study Guide* has been a collaborative effort. I am most grateful for the goodwill and assistance given to me by all those involved in its creation.

Some of the companies and organizations that contributed include the Associated Locksmiths of America (ALOA), the International Foundation for Protection Officers, and the International Association of Home Safety and Security Professionals.

I'd also like to thank my good friend Joanne Goodwine for pushing me to finish this book, as well as my son Michael and sister Merlynn for always being there for me. Most of all I want to thank my McGraw-Hill editor, Joy Bramble Oehlkers, for her tremendous patience and guidance, without which this book would not have been published.

LOCKSMITH AND SECURITY PROFESSIONALS' EXAM STUDY GUIDE

Chapter 1
LOCK HISTORY

This chapter traces the development of the lock from earliest times to the present, focusing on the most important models. Every locksmith should be familiar with these models because they form the building blocks for all other locks. Many of the lock types and construction principles mentioned here are looked at in more detail in later chapters. This chapter is a quick overview to help you better understand and appreciate the world of locks.

Who Invented the Lock?

The earliest locks may no longer be around, and there may be no written records of them. How likely it is for old locks to be found depends on the materials they were made from, and on the climate and various geological conditions they were subjected to over the years. Evidence exists to suggest that different civilizations probably developed the lock independently of each other. The Egyptians, Romans, and Greeks are credited with inventing the oldest known types of locks.

The oldest known lock was found in 1842, in the ruins of Emperor Sargon II's palace in Khorsabad, Persia. The ancient Egyptian lock was dated to be about 4000 years old. It relied on the same pin-tumbler principle used by many of today's most popular locks.

The Egyptian lock consisted of three basic parts: a wood crossbeam, a vertical beam with tumblers, and a large wood key. The crossbeam ran horizontally across the inside of the door and was held in place by two vertically mounted wooden staples. Part of the length of the crossbeam was hollowed out, and the vertical beam intersected it along that hollowed-out side. The vertical beam contained metal tumblers that locked the two pieces of wood together. Near the tumbler edge of the door, a hole—large enough for someone to insert the key and an arm—was accessible from outside the door. The spoon-shaped key was about 14 inches to 2 feet long with pegs sticking out of one end. After the key was inserted in the keyhole (or "armhole"), it was pushed into the hollowed-out part of the crossbeam until its pegs were aligned with their corresponding tumblers. The right key allowed all the tumblers to be lifted into a position between the crossbeam and the vertical beam, so the pins no longer obstructed the movement of the crossbeam. Then, the crossbeam (bolt) could be pulled into the open position.

Greece

Most early Greek doors pivoted at the center and were secured with rope tied in intricate knots. The cleverly tied knots, along with beliefs about being cursed for tampering with them, provided some security. When more security was needed, doors were secured by bolts from the inside. In the few cases where locks were used, they were primitive and easy to defeat. The Greek locks used a notched boltwork and were operated by inserting the blade of an iron sickle-shaped key, about a foot long, in a key slot and twisting it 180° to work the bolt. They could be defeated just by trying a few different-sized keys.

In about 850 B.C., the Greek poet Homer described that Greek lock in his epic, *The Odyssey*:

"She went upstairs and got the store room key, which was made of bronze and had a handle of ivory; she then went with her maidens into the store room at the end of the house, where her husband's treasures of gold, bronze, and wrought iron were kept.... She loosed the strap from the handle of the door, put in the key, and drove it straight home to shoot back the bolts that held the doors."

Like the Greeks, the Romans used notched boltwork. But the Romans improved on the lock design in many ways, such as by putting the boltwork in an iron case and using keys of iron or bronze. Because iron rusts and corrodes, few early Roman locks are in existence. But a lot of the keys are around. Often, the keys were ornately designed to be worn as jewelry, either as finger rings or as necklaces using string (because togas didn't have pockets).

Two of the most important innovations of the Roman locks were the spring-loaded bolt and the use of wards on the case. The extensive commerce during the time of Julius Caesar led to a great demand for locks among the many wealthy merchants and politicians. The type of lock used by the Romans, the warded bit-key lock, is still being used today in many older homes. Because the lock provides so little security, typically it's found on interior doors, such as closets, and sometimes bedrooms.

The Romans are sometimes credited with inventing the padlock, but that's controversial. Evidence exists that the Chinese may have independently invented it before or at about the same time.

The demand for locks declined after the fall of Rome in the fifth century because people had little property to protect. The few locks used during the period were specially ordered for nobility and the handful of wealthy merchants.

Europe

During the Middle Ages, metal workers in England, Germany, and France continued to make warded locks, with no significant security changes. They focused on making elaborate ornately designed cases and keys. Locks became works of art.

Keys were made that could move about a post and shift the position of a movable bar (the locking bolt). The first obstacles to unauthorized use of the lock were internal wards. Medieval and renaissance craftsmen improved on the warded lock by using many interlocking wards and more complicated keys. But many of the wards could easily be bypassed.

In 1767, the treatise, *The Art of the Locksmith*, was published in France. It described examples of the lever tumbler lock. The inventor of the lock is unknown. As locksmithing advanced, locks were designed with multiple levers, each of which had to be lifted and properly aligned before the bolt could move to the unlocked position.

In the fourteenth century, the locksmiths' guilds came into prominence. They required journeymen locksmiths to create and submit a working lock and key to the guild before being accepted as a master locksmith. The locks and keys weren't made to be installed, but to be displayed in the guild hall. The guilds' work resulted in some beautiful locks and keys. The problem with the locksmith guilds was they gained too much control over locksmiths, including the regulating of techniques and prices. The guilds became corrupt and didn't encourage technological advances. Few significant security innovations were made because of the locksmiths' guilds. The innovations included such things as false and hidden keyholes. A fish-shaped lock, for instance, might have the keyhole hidden behind a fin.

England

Little progress was made in lock security until the eighteenth century. Incentive was given in the form of cash awards and honors to those who could pick open newer and more complex locks. That resulted in more secure lock designs. In the forefront of lock designing were three Englishmen: Robert Barron, Joseph Bramah, and Jeremiah Chubb.

The first major improvement over warded locks was patented in England in 1778 by Robert Barron. He added the tumbler principle to wards for increased security. Barron's double-acting lever tumbler lock was more secure than other locks during that time and remains today the basic design for lever tumbler locks. Like other lever tumbler locks, Barron's used wards. But, Barron also used a series of lever tumblers, each of which was acted on by a separate step of the key. If any tumbler wasn't raised to the right height by the key, its contact with a bolt stump would obstruct bolt movement. Barron's lock corrected the shortcomings of earlier lever-tumbler locks, which could easily be circumvented by any key or instrument thin enough to bypass the wards. Barron added up to six of these double-lever actions to his lock and thought it was virtually impossible to open it except by the proper key. He soon found out differently.

Another Englishman, Joseph Bramah, wrote *A Dissertation on the Construction of Locks*, which exposed the many weaknesses of existing so-called thiefproof locks. He pointed out that many of them could be picked by a good specialist or a criminal with some training in locks and keys. Bramah admitted that Barron's lock had many good points, but he also revealed its major fault: the levers, when in the locked position, gave away the lock's secret. The levers had uneven edges at the bottom; thus, a key coated with wax could be inserted into the lock and a new key could be made by filing where the wax had been pressed down or scraped away. Several tries could create a key that matched the lock. Bramah pointed out that the bottom edges of the levers showed exactly the depths to which the new key should be cut to clear the bolt. Bramah suggested that the lever bottoms should be cut unevenly. Then, only a master locksmith should be able to open it.

Using those guidelines, Bramah patented a barrel-shaped lock in 1798 that employed multiple sliders around the lock, which were to be aligned with corresponding notches around the barrel of its key. The notches on the key were of varying heights. When the right key was pushed into the lock, all the notches lined up with the sliders, allowing the barrel to rotate to the unlocked position. It was the first to use the rotating element in the lock itself.

During this period, burglary was a major problem. After the Portsmouth, England, dockyard was burglarized in 1817, the British Crown offered a reward to anyone who could make an unpickable lock. A year later, Jeremiah Chubb patented his lock and won the prize money.

Jeremiah Chubb's detector lock was a four-lever tumbler rim lock that used a barrel key. It had many improvements over Barron's lock. One of the improvements was a metal "curtain" that fell across the keyhole when the mechanism began to turn, making the lock hard to pick. Chubb's lock also added a detector lever that indicated whether the lock had been tampered with. A pick or an improperly cut key would raise one of the levers too high for the bolt gate. That movement engaged a pin that locked the detector lever. The lever could be cleared by turning the correct key backward, and then forward.

Chubb's lock got much attention. It was recorded that a convict who had been a lockmaker was on board one of the prison ships at Portsmouth Dockyard and said he had easily picked open some of the best locks and he could easily pick open Chubb's detector lock. The convict was given one of the locks and all the tools that he asked for, including key blanks fitted to the drill pin of the lock. As incentive to pick open the lock, Mr. Chubb offered the convict a reward of £100, and the government offered a free pardon if he succeeded. After trying for several months to pick the lock, the convict gave up. He said that Chubb's lock was the most secure lock he had ever met with and that it was impossible for anyone to pick or open it with false instruments. The lock was improved on by Jeremiah's brother, Charles Chubb, and Charles's son, John Chubb, in several ways, including the addition of two levers and false notches on the levers.

The lock was considered unpickable until it was picked open in 1851 at the International Industrial Exhibition in London by an American locksmith named Alfred C. Hobbs. At that event, Hobbs picked open both the Bramah and the Chubb locks in less than half an hour.

America

During America's early years, England had a policy against its skilled artisans leaving the country. This was to keep the artisans from running off and starting competing foreign companies. Locks made by early American locksmiths didn't sell well. In the mid 1700s, few colonists used door locks, and most that were used were copies of European models. More often, Americans used lock bolts mounted on the inside of the door that could be opened from the outside by a latchstring, hence, the phrase, "the latchstring's always out." At night, the string would be pulled inside, "locking" the door. Of course, someone had to be inside to release the bolt. An empty house was left unlocked. As the country settled, industry progressed and theft increased, creating a rising demand for more and better locks. American locksmiths soon greatly improved on the English locks and were making some of the most innovative locks in the world. Before 1920, American lock makers patented about 3000 different locking devices.

In 1805, an American physician, Abraham O. Stansbury, was granted an English patent for a pin tumbler lock that was based on the principles of both the Egyptian and Bramah locks. Two years later, the design was granted the first lock patent by the U.S. Patent and Trademark Office. Stansbury's lock used segmented pins that automatically relocked when any tumbler was pushed too far. The double-acting pin tumbler lock was never manufactured for sale.

In 1836, a New Jersey locksmith, Solomon Andrews, developed a lock that had adjustable tumblers and keys, which allowed the owner to rekey the lock anytime. Because the key could also be modified, there was no need to use a new key to operate a rekeyed lock. But few homeowners used the lock because rekeying it required dexterity, practice, and skill. The lock was of more interest to banks and businesses.

In the 1850s, two inventors—Andrews and Newell—were granted patents on an important new feature: removable tumblers that could be disassembled and scrambled. The keys had interchangeable bits that matched the various tumbler arrangements. After locking up for the night, a prudent owner would scramble the key bits. Even if a thief got possession of the key, stumbling onto the right combination would take hours. In addition to removable tumblers, this lock featured a double set of internal levers.

Newell was so proud of this lock, he offered a reward of $500 to anyone who could open it. A master mechanic took him up on the offer and collected the money. This experience convinced Newell that the only secure lock would have its internals sealed off from view. Ultimately, the sealed locks appeared on bank safes in the form of combination locks.

Until the time of Alfred C. Hobbs, who picked the famed English locks with ease, locks were opened by making a series of false keys. If the series was complete, one of the false keys would match the original. Of course, this procedure took time. Thousands of hours might pass before the right combination was found. Hobbs depended on manual dexterity. He applied pressure on the bolt, while manipulating one lever at a time with a small pick inserted through the keyhole. As each lever tumbler unlatched, the bolt moved a hundredth of an inch or so.

Hobbs patented what he called "Protector" locks, but they weren't invincible either. In 1854, one of Chubb's locksmiths used special tools to pick open one of Hobbs's locks.

Until the early nineteenth century, locks were made by hand. Each locksmith had his own ideas about the type of mechanism—the number of lever tumblers, wards, and internal cams to put into a given lock. Keys contained the same individuality. A lock could have 20 levers and weigh as much as 5 pounds.

In 1844, Linus Yale, Sr., of Middletown, Connecticut, patented his "Quadruplex" bank lock, which incorporated a combination of ancient Egyptian design features and mechanical principles of the Bramah and Stansbury locks. The *Quadruplex* had a cylinder subassembly that denied access to the lock bolt. In 1848, Yale patented another pin tumbler design based on the Egyptian and the Bramah locks. His early models had the tumblers built into the case of the lock and had a round fluted key. His son, Linus Yale, Jr., improved on the lock design and is credited with inventing the modern pin tumbler lock.

Arguably, the most important modern lock development is the Yale Mortise Cylinder Lock, U.S. patent 48,475, issued on June 27, 1865, to Linus Yale, Jr. This lock turned the lock-making industry upside down and established a new standard. Yale, Jr.'s lock could not only easily be rekeyed, but it also provided a high level of security; it could easily be mass produced; and it could be used on doors of various thicknesses. Linus Yale, Jr.'s lock design meant that keys no longer had to pass through the thickness of the door to reach the tumblers or bolt mechanism, which allowed the keys to be made thinner and smaller. (Linus Yale, Jr.'s first pin tumbler locks used a flat steel key, rather than the paracentric cylinder type often used today.)

Since 1865, few major changes have occurred to the basic design of mechanical lock cylinders. Most cylinder refinements since that time have been limited to using unique keyways (along with corresponding shaped keys), adding tumblers, varying tumbler positions, varying tumbler sizes and shapes, and combining two or more basic types of internal construction—such as the use of both pin tumblers and wards. Most major changes in lock design have centered around the shape and installation methods of the lock.

In 1916, Samuel Segal, a former New York City police officer, invented the jimmyproof rim lock (or "interlocking deadbolt"). The surface-mounted lock has vertical bolts that interlock with "eye-loops" of its strike, locking the two parts together in such a way that you would have to break the lock to pry them apart.

In 1920, Frank E. Best received his first patent for an interchangeable core lock. It allows you to rekey a lock just by using a control key and switching cores. The core was made to fit into padlocks, mortise cylinders, deadbolts, key-in-knobs, and other types of locks.

In 1833, three brothers—the Blake brothers—were granted a patent for a unique door latch that had two connecting doorknobs. It was installed by boring two connecting holes. The larger hole, which was drilled through the door face, was for the knob mechanism. The smaller hole, which was drilled through the door edge, was for the latch. The big difference between their latch and others of their time was that all the door locks were installed by being surface mounted to the inside surface of a door. In 1834, the brothers formed the Blake Brothers Lock Company to produce and sell their unusual latch. At that time, the brothers probably never imagined that nearly 100 years later, their creation would be used to revolutionize lock designs.

In 1928, Walter Schlage patented a cylindrical lock that incorporates a locking mechanism between the two knobs. Schlage's was the first knob-type lock to have mass appeal. Today, key-in-knob locks are commonplace.

In 1933, Chicago Lock Company introduced its tubular key lock, called the Chicago Ace Lock, which was based on the pin tumbler principle, but used a circular keyway. The odd keyway made it hard to pick open without using special tools. For a long time, many locksmiths referred to all tubular key locks as "Ace Locks," not realizing that was only a brand name. Today, the lock is made by many manufacturers and is used on vending machines, in padlocks, and for bicycle locks.

A recent innovation in high-security mechanical locks came in 1967, with the introduction of the Medeco high-security cylinder. The cylinder, made by Roy C. Spain and his team, used chisel-pointed rotating pins and restricted angularly bitted keys that made picking and impressing harder. To open the lock, a key not only had to simultaneously lift each pin to the proper height, but it also had to rotate each one to the proper position to allow a sidebar to retract. The name "Medeco" was based on the first two letters of each word of the name "Mechanical Development Company." The Medeco Security Lock was the largest and most talked about high-security lock. In the early 1970s, the company offered a reward for anyone who could pick open one, two, or three of its cylinders within a set amount of time. In 1972, Bob McDermott, a New York City police detective, picked one open in time and collected the reward. But that feat didn't slow the demand for Medeco locks. Much of the general public never heard about the contest and still considered Medeco locks to be invincible. In 1986, Medeco won a patent infringement lawsuit against a locksmith who was making copies of Medeco keys. That ruling stopped most other locksmiths from making the keys without signing up with Medeco. The patent for the original Medeco key blank ran out, and now anyone can make keys for those cylinders. In 1988, the company received a new patent for its "biaxial" key blanks. The big difference is that the biaxial brand gave Medeco a new patent (which can be helpful for preventing unauthorized key duplication). The company's newest lock is the Medeco.

Early American Lock Companies

In 1832, the English lock maker, Stephen G. Bucknall, became the first trunk and cabinet lock manufacturer in America. Bucknall made about 100 cabinet lock models, but he didn't sell many. After his company folded, Bucknall went to work for Lewis, McKee & Company. William E. McKee was a major investor in the company. In 1835, Bucknall left and received financial assistance from McKee to form the first trunk lock company in America, called the Bucknall, McKee Company. A couple years later, Bucknall sold the business and went to work for North & Stanley Company.

One of the greatest successes and failures of the American lock industry was the Eagle Lock Company, formed in 1854. It was the result of a merger between the James Terry Company and Lewis Lock Company. Eagle Lock had money to burn and was quick to buy out its competitors, such as American Lock Company, Gaylord, and Eccentric Lock Company. In 1922, the company had 1800 employees and several large warehouses. In 1961, the company introduced a popular line of locks and cylinders called "Supr-Security." Their line was highly pick-resistant, and their keys couldn't be duplicated on standard key machines. The company was sold to people who didn't know a lot about locks, and many bad decisions were made, which sent the company on a downward path. Profits were being siphoned off without considering the long-term needs of the business. Top management quit. By the early 1970s, the company was barely holding on. In 1973, a businessperson bought it from Penn-Akron Corporation, which had gone bankrupt. In 1974, Eagle Lock lost a bid to the Lori Corporation for a large order of cylinders for the U.S. Postal Service. A few months later, the company folded. The Lori Corporation bought the Eagle Lock equipment at a public auction, and the Eagle Lock plants were burned, ending 122 years of lock making.

A Brief History of Automotive Locks in the United States

While the car has been with us since the beginning of the twentieth century, the automobile lock was adopted slowly. By the late 1920s, however, nearly every vehicle had an ignition lock, and closed cars had door locks as well. Current models can be secured with half a dozen locks. This section gives an overview of the history of automobile locks in the U.S. The information is especially useful when you're working on older vehicles. (For in-depth information on servicing and opening all types of automobiles, see Chapter 8.)

1935 General Motors began using sidebar locks. There was only one keyway, with 6 cuts and 4 depths.

1959 Chrysler began using sidebar trunk locks.

1966 Chrysler stopped using sidebar trunk locks, and began using pin tumbler locks, which were the same size as the door locks.

1967 General Motors added a 5th depth to their codes and introduced two new key blanks—P1098A and S1098B.

1968 General Motors introduced two new key blanks: P1098C and S1098D.

1969 General Motors began using steering column–mounted ignition locks and introduced two new key blanks—P1098E and S1098H.

1970   American Motors, Chrysler, and Ford began using column-mounted ignition locks. General Motors stopped putting codes on door locks and introduced two new blanks—P1098J and S1098K.

1972   Chrysler began using General Motors' Saginaw tilt steering columns with sidebar ignition locks.

1973   Chrysler began using trunk locks retained by large nuts.

1974   General Motors began keying locks, so the primary key fit the ignition lock only and all other locks on the vehicle used the secondary key.

1977   Ford stopped putting codes on door locks.

1978   General Motors stopped putting codes on glove compartment locks.

1979   General Motors changed ignition locks from being spring tab retained to screw retained.

1979   Ford began using fixed pawls on door locks.

1980   General Motors began using fixed pawls on door locks.

1981   Ford began keying locks, so the primary key fit the ignition and the secondary key fit all other locks. The company stopped putting codes on glove compartment locks.

1983   American Motors began making the Alliance, using X116 ignition key and X92 door key, both previously known as foreign auto keys.

1985   Ford began using sidebar ignition locks and wafer tumbler door locks, which worked with Ford 10-cut keys. Primary keys operate doors and ignitions; secondary keys operate glove compartments and trunks.

1986   The Vehicle Antitheft System (VATS) was introduced by General Motors on the 1986 Corvette.

1989   Chrysler introduced its double-sided wafer tumbler locks.

1992   General Motors introduced the Mechanical Antitheft System (MATS) in its full-sized rear-wheel-drive cars—Oldsmobile Custom Cruiser Station Wagon, Buick Headmaster, and Chevrolet Caprice.

1996   The Passive Antitheft System (PATS), a radio frequency identification system, was introduced on the 1996 Ford Taurus and Mercury Sable.

Chapter Quiz

1. Who is credited with inventing the same pin tumbler principle used by many of today's most popular locks?

   A. The Romans

   B. The Egyptians

   C. The Chinese

   D. The Africans

2. Who patented a barrel-shaped lock in 1798?

   A. Joseph Bramah

   B. Joe Master

   C. Linus Yale

   D. John Kwikset

3. Whose lock consisted of three basic parts: a wood crossbeam, a vertical beam with tumblers, and a large wood key?

   A. The Romans

   B. The Chinese

   C. The Egyptians

   D. The Africans

4. Who wrote *A Dissertation on the Construction of Locks*?

   A. Joseph Bramah

   B. Linus Yale

   C. Joe Master

   D. John Kwikset

5. The first major improvement over warded locks was patented in England in 1778 by Robert Barron.

   A. True   **B.** False

6. Two of the most important innovations of the Roman locks were the spring-loaded bolt and the use of wards on the case.

   A. True   **B.** False

7. Jeremiah Chubb's detector lock was a four-lever tumbler rim lock that used a barrel key.

   A. True   **B.** False

8. In 1844, Linus Yale, Sr., of Middletown, Connecticut, patented his "Quadruplex" bank lock, which incorporated a combination of ancient Egyptian design features and mechanical principles of the Bramah and Stansbury locks.

   A. True   **B.** False

9. Like the Greeks, the Romans used notched boltwork. But the Romans improved on the lock design in many ways, such as putting the boltwork in an iron case and using keys of iron or bronze.

   A. True   **B.** False

10. To impression a lock, you need to choose the right blank.

    A. True   **B.** False

11. Hobbs patented what he called "Protector" locks.

    A. True   **B.** False

12. In 1916, Samuel Segal, a former New York City police officer, invented the jimmyproof rim lock (or interlocking deadbolt).

    A. True   **B.** False

13. The Vehicle Antitheft System (VATS) was introduced by General Motors on the 1998 Corvette.

    A. True   **B.** False

14. The Passive Antitheft System (PATS), a radio frequency identification system, was introduced on the 2008 Ford Taurus and the Mercury Sable.

    A. True   **B.** False

15. In 2008, General Motors introduced the Mechanical Antitheft System (MATS) in its full-sized rear-wheel-drive cars—Oldsmobile Custom Cruiser Station Wagon, Buick Headmaster, and Chevrolet Caprice.

    A. True   **B.** False

16. In 2007, General Motors began using steering column–mounted ignition locks and introduced two new key blanks—P1098E and S1098H.

    A. True   **B.** False

Chapter 2

PRIVATE SECURITY AND LOSS PREVENTION

Contemporary England

Repeated attempts were made to improve protection and justice in England. Each king faced increasingly serious problems from crime and cries from the citizenry for help. As England colonized many parts of the world, and as trade and commercial pursuits brought many people into the cities, urban problems and high crime rates continued. Unsatisfied by the protection they were receiving, merchants hired private security forces to protect their businesses.

Peel's Reforms

The Metropolitan Police Act, which was the birth of modern policing, came about through the efforts of Sir Robert Peel in 1829. Peel's innovative ideas were accepted by Parliament, and he was chosen to implement an act that established a full-time, unarmed police force with the main purpose of patrolling London. Peel is also credited with reforming the criminal law by limiting its scope and abolishing the death penalty for more than 100 crimes. The hope was that such reforms would gain public support and respect for the police. Peel was selective in his choice of hiring, and he emphasized professional training. His reforms are applicable today, and they include crime prevention, the strategic deployment of police according to time and location, record keeping, and distribution of crime news.

The Growth of Policing

During the mid 1800s, a turning point occurred in both law enforcement and private security in England and America. Several major cities, including New York, Philadelphia, and San Francisco, organized police forces—many of which were modeled after the London Metropolitan Police. But corruption was widespread. Many urban police forces received large boosts in personnel and resources to combat the growing militancy of the labor unions in the late 1800s and early 1900s. Many of the large urban police forces were formed originally as strikebreakers. In 1864, the U.S. Treasury had already established an investigative unit. As in England, an increase in paid police officers didn't eliminate the need for private security.

Early America

The Europeans who colonized North America brought the legal and security heritage of their mother countries. The watchman system and collective responses were popular. A central fortification in largely populated areas provided increased security from crime. As communities grew, the office of sheriff took hold in the south, whereas the functions of constable and watchman were the norm in the northeast. The sheriff's duties included catching criminals, serving subpoenas, and collecting taxes. Constables were responsible for keeping the peace, bringing suspects and witnesses to court, and eliminating health hazards. As in England, the watch system was inefficient, and those convicted of minor crimes were sentenced to severe time on the watch.

The Growth of Private Security in America

In 1850, after becoming Chicago's first police detective, Allan Pinkerton, a cooper, started a detective agency. Police departments were limited by jurisdictions. But the private security companies could cross state lines to apprehend fleeing criminals. Pinkerton and other private security companies became famous for catching criminals who crossed jurisdictions. Today, Pinkerton is a subsidiary of Securitas, based in Stockholm, Sweden.

In 1852, William Fargo started Wells, Fargo & Company to ensure the safe transportation of valuables. Today, Wells Fargo is a division of Burns International Services Corporation, a subsidiary of Securitas.

William Burns was another security entrepreneur. First, Burns was a Secret Service agent who directed the Bureau of Investigation, which preceded the FBI. In 1909, he opened the William J. Burns Detective Agency, which became an arm of the American Bankers Association. Today, Burns International Services Corporation is a subsidiary of Securitas.

In 1859, Washington P. Brink also took advantage of the need for safe transportation of valuable freight, package, and payroll delivery. As cargo became more valuable through the years, Brink's service required increased protection. In 1917, following the killing of two Brink's guards during a robbery, the armored truck was initiated. Today, Brink's, Inc., a subsidiary of the Pittston Company, is the world's largest provider of secure transportation services. It also does considerable business monitoring home-alarm systems.

Another major figure in the history of private security in the United States is Edwin Holmes. He pioneered the electronic alarm business. During 1858, Holmes had a hard time convincing people that an alarm would sound on the second floor of a home when a door or window was opened on the first floor. He carried a model of his electronic alarm system door-to-door. Holmes installed the first of his alarm systems in Boston on February 21, 1858. Sales of his system soon soared, and the first central monitoring station was formed. Holmes Protection Group, Inc., was purchased by ADT Security Services, Inc.

Since 1874, ADT Security Services, Inc., has been a leader in electronic alarm services. The company was known originally as American District Telegraph. Today ADT has acquired many security companies. It is a unit of Tyco Fire and Security Services and is the largest provider of electronic security services—serving nearly 3 million commercial, federal, and residential customers throughout North America and the United Kingdom.

Another leader in the private security industry is the Wackenhut Corporation, which provides correctional and human resources services. The company was founded in 1954 by George Wackenhut, a former FBI agent, and today, it has operations throughout all 50 states in the United States.

Railroads and Labor Unions

The history of private security in the United States is largely related to the growth of railroads and labor unions. Railroads were important for providing the East-West link that allowed the settling of the American Frontier, but the powerful businesses used their domination of transportation to control several businesses, such as coal and kerosene. Farmers had to pay high fees to transport their products by train. These monopolistic practices created much hostility. Citizens applauded when Jessie James and other criminals robbed trains. Because of jurisdictional boundaries, railroads couldn't rely on public police protection. Many states passed laws allowing railroads to have their own private security forces, with full arrest powers and the authority to arrest criminals who crossed jurisdictions. By 1914, there were 14,000 railroad police. During World War I, railroad police were deputized by the federal government to ensure protection of this vital means of transportation.

The growth of labor unions during the nineteenth century resulted in an increased need for strikebreakers for large businesses. This was a costly business. A bloody confrontation between Pinkerton men and workers at the Carnegie steel plant in Homestead, Pennsylvania, resulted in eight deaths (three security men and five workers). Pinkerton's security forces withdrew. Then, the plant was occupied by federal troops.

As a result, the Homestead disaster and anti-Pinkertonism laws were passed to restrict private security. Local and state police forces then emerged to deal with strikers. Later, the Ford Motor Company and other businesses were involved in bloody confrontations. Henry Ford had a private security force of about 3,500 people, who were helped by various community groups, such as the Knights of Dearborn and the Legionnaires. Media coverage of the confrontations gave a negative impression of private security. Before World War II, the Roosevelt administration, labor unions, and the American Civil Liberties Union (ACLU) forced corporate management to shift its philosophy to a softer approach.

World Wars I and II

World Wars I and II brought about an increased need for protection in the United States. Sabotage and espionage were big threats. Major industries and transportation systems needed expanded and improved security. The social and political climate in the early twentieth century reflected urban problems, labor unrest, and worldwide nationalism. World War I compounded these concerns and people's fears. A combination of the war, Prohibition, labor unrest, and the Great Depression overtaxed public police forces. Private security helped to fill the void.

By the late 1930s, Europe was at war again, and the Japanese were expanding in the Far East. In 1941, a surprise Japanese bombing of the Pacific fleet at Pearl Harbor pushed the United States into the war, and security again became a major concern for people. Protection of vital industries became critical, causing the United States to bring plant security personnel into the army as auxiliaries to military police. By the end of the war, over 200,000 of these security workers had been sworn in.

Twenty-First Century Security

The 1990s brought the first bombing of the World Trade Center and the bombing of the Murrah Federal Building in Oklahoma City, war with Iraq, crimes over the Internet, increased value of proprietary information, and more violence in the workplace, all leading to an increased need for private security. Current concerns over security notwithstanding, security increasingly has been an issue throughout American history. From the labor struggles of the early twentieth century through the necessary security of World War II and the Cold War to the increased crime of the sixties and seventies, Americans have sought ways to protect their property and stay safe.

Brink's Home Security was founded in 1983 as an affiliate of Brink's, Inc.—the world's largest provider of secure transportation services. Its headquarters are in Irving, Texas, where the National Service Center (NSC) provides alarm monitoring and other customer services. Today, Brink's Home Security monitors home security systems for more than 1 million customers in over 200 markets in the United States and Canada.

Early in the twenty-first century, on September 11, 2001, terrorists hijacked airplanes and attacked the World Trade Center and the Pentagon, resulting in about 3000 deaths. Such a bold surprise attack shows how challenging security is now and how it requires a new way of thinking. Security professionals need to consider how to prevent not only a variety of accidents, disasters, crimes, and fire, but also terrorism and bioterrorism. These days, it's critical for loss-prevention officers to have improved education and training, so they can face current challenges professionally, and with creativity and imagination.

Chapter Quiz

1. Who is credited with reforming the criminal law by limiting its scope and abolishing the death penalty for more than 100 crimes?

   A. D. Templeton

   B. Robert Pinkerton

   C. Sir Robert Peel

2. In 1850, after becoming Chicago's first police detective, who started a detective agency?

   A. Sir Robert Peel

   B. Allan Pinkerton

   C. George Wackenhut

3. Brink's Home Security was founded in 1983, as an affiliate of Brink's, Inc.

   A. True   **B.** False

4. The growth of labor unions during the nineteenth century resulted in an increased need for strikebreakers for large businesses.

   A. True   **B.** False

5. The Homestead disaster and anti-Pinkertonism laws were passed to restrict private security.

   A. True   **B.** False

6. In 1850, Allan Pinkerton opened a detective agency in the United States.

   A. True   **B.** False

7. The history of private security in the United States is largely related to the growth of railroads and labor unions.

   A. True   **B.** False

8. On September 11, 2003, terrorists hijacked airplanes and attacked the World Trade Center and the Pentagon, resulting in about 3000 deaths.

   A. True   **B.** False

9. Since 1874, ADT Security Services, Inc., has been a leader in electronic alarm services.

   A. True   **B.** False

10. What company was founded in 1954 by George Wackenhut, a former FBI agent, and today, has operations throughout the 50 U.S. states?

    _____

11. The Metropolitan Police Act, which was the birth of modern policing, came about through the efforts of Sir Robert Peel in 1829.

    A. True   **B.** False

12. Pinkerton is now a subsidiary of Securitas, based in Stockholm, Sweden.

    A. True   **B.** False

13. Wells Fargo is now a division of Burns International Services Corporation, a subsidiary of Securitas.

    A. True   **B.** False

14. World Wars I and II brought about an increased need for protection in the United States. Sabotage and espionage were big threats.

    A. True   **B.** False

15. Since 1874, ADT Security Services, Inc. has been a leader in electronic alarm services. The company was known originally as American District Telegraph.

    A. True   **B.** False

Chapter 3
SECURING DOORS

Whenever I'm asked about building security, the questions primarily concern door locks. Rarely does anyone think to ask about making the doors themselves stronger. I've been in homes where people were asking me to recommend a lock for a door that looked as though it would fall down if anyone knocked on it too hard.

It's important to know that a *door lock* is just a device that fastens a door to one or more sides of a door frame. Using a good lock on a thin-paneled door or on a door with weak hinges is like using a heavy-duty padlock to secure a paper chain. Before worrying about a good lock, be sure you have a strong door and frame.

How Intruders Can Open Doors

Burglars aren't invited guests and shouldn't be allowed to enter buildings as though they were. Be aware of burglars' secrets for getting past doors, and what you can do to keep them out.

Removing Hinges

If a door's hinges can be seen from the exterior side, a burglar may be able to remove them and open the door without touching the lock. Most door hinges consist of two metal leaves (or *plates*)—each with "knuckles" on one edge—and a pin that fits vertically through the knuckles when they're aligned and holds the leaves together. The hinge pins often can be pulled out with little difficulty, and the door then becomes disconnected from the door frame. Burglars who remove a door in that way can place the door back in its hinges on the way out, and how they got in may never be determined (which may result in an unpaid insurance claim). One way to prevent this type of entry is to use hinges with *nonremovable pins*—pins that are either welded in place or secured by a set screw or retaining pin.

An alternative to replacing the door hinges is to install *hinge enforcers*, which are small metal devices that attach to the hinge and the door frame to block the door's removal, even when the hinge pins have been removed. A package of hinge enforcers costs less than $20.

Prying Off Stop Molding

If a door's hinges can't be seen from the outside, your next concern should be the door's stop molding. *Stop moldings* are the protruding strips (usually about ½-inch thick) that are installed on three sides of a door frame—the lock side, the hinge side, and the header (top). They stop the door from swinging too far when someone is closing it. Depending on which way the door swings, a person standing outside the door will be able to see either the hinges or the stop molding.

Some stop moldings are simply thin wooden strips tacked to the frame, and they can be easily pried off. By removing the lock side strip, an intruder exposes the bolt and makes it easier to attack the lock. To solve the problem, the stop molding can be removed and reinstalled using wood glue and nails, so it can't easily be pried off. When you buy or make a new door, be sure its stop molding is milled as an integral part of the jamb.

Knocking Off the Key-in-Knob Lock

Most building doors have a key-in-knob lock. Such locks are convenient, because they lock a door simply by pulling the door closed. The problem with such locks is an intruder can bypass them simply by using a hammer to knock off the exterior knob, and then by using a screwdriver to unlock the door. That's why every exterior key-in-knob lock should be installed with a deadbolt lock.

Kicking Doors Down

One of the most common ways burglars get through doors is by kicking them. If either the strike plate on the door jamb or the lock edge of the door is weak, a strong kick can knock the door open. Short of getting a new door, the best way to solve a weak door problem is to install door reinforcers, which usually cost less than $20 each.

One type of door reinforcer is a U-shaped metal unit designed to wrap around the door edge under the lock. Designs are available for doors with one or two locks. To install this type of reinforcer, first remove the locks from the door. Then, position the unit, so the lock holes are fully exposed, and then screw it firmly into place. Next, install the locks.

Weak door frames can also be strengthened. A popular reinforcer for door frames is the *high-security strike box*, a heavy-gauge steel box with long screws or rods that protrude through the door jamb and into a wall stud. The strike box is stronger than the more commonly used thin, flat strike plates that are fastened only to the jamb, using small wood screws.

Sliding Glass Doors

A sliding glass door (sometimes called a *patio* door) usually consists of two glass panels (or "sashes") that slide along tracks. Doors of this type are especially vulnerable because their frames and locks are weak. A sliding glass door can be forced open by prying the sliding panel away from the door frame. That entry technique can be thwarted by sliding door barriers.

Another way burglars can defeat sliding glass doors is by using a pry bar to pry the sliding sash out of its lower track. Installation of screws or antilift plates at the top of the door can thwart this entry technique. A package of such plates costs less than $10.

Garage Doors

Don't ignore garage doors. Burglars know that a typical garage contains cars, bikes, lawn mowers, tools, and other easy-to-sell items. And a garage that's attached to a home usually provides easy access to the home. The most secure main garage doors are made of steel, require an automatic door opener, and have no glass or thin panels.

How to Reinforce Garage Door Panels with Angel Iron

1. Using ¾-inch by ¾-inch angel iron that nearly spans the width of the garage door, position the angel iron, so it crosses the horizontal center of a row of panels.
2. Mark locations for screw holes along the bar about every 2 inches, if possible. Remove the bar from the door, and then drill screw holes through the bar at the marked points.
3. Place the bar back into position on the door and use an awl to punch starter holes into the door. Then, screw the angel iron into place.

Panels made of any material weaken a door, but glass panels are an especially poor feature in garage doors. They can be broken easily, and they let a burglar see what's in the garage. In addition to securing the main door of a garage, any door that allows passage from the garage to the home should be secured. That "inside" door should be as secure as any exterior door. Burglars who are able to drive into the garage and enter the home through the garage entrance will be unseen while they load the stolen possessions into their car.

Choosing a New Door

In most homes, the style of door depends on the structure's architecture. Typically, people want doors that complement their home's design. Usually, an exterior door is connected to its frame by metal hinges on one side and a lock on the other. The frame consists of various sections: a head jamb (along the top), two side jambs, stop molding along the top and sides, and a sill or threshold (along the bottom). Although the door and frame don't have to be made of the same material, they usually are. Commonly used materials include steel, wood, aluminum, fiberglass, polyvinyl chloride (PVC) plastic, and glass.

How to Replace an Exterior Door

Before prehung doors were introduced, the parts of the frame had to be cut, assembled, glued, and nailed together, and then the door had to be hung on the frame (using hinges). A prehung door unit comes already assembled on a frame, and it's ready to position and fasten in place. You can replace an exterior door with a prehung exterior door by doing the following:

1. Pry off the interior trim from around the four sides of the door. You have to work a pry bar along the entire length of each strip of trim. (Be careful not to scratch the wall around the door.)
2. Using a punch and mallet, remove the pins from each of the hinges on the door. Then, remove the door (you may need a helper).
3. Use a saw to cut the threshold (the button of the door frame) into three sections. Then, use a pry bar to remove each section of the threshold from the sill.
4. Pry the side jambs from the studs and the head jamb from the header. Take out any shims or nails that may be sticking out of the sill, studs, or header.
5. If necessary, nail plywood strips to the studs and header to make the dimensions of the rough opening about ½ inch larger than the overall size of the new door frame.
6. Run two parallel beads of caulk along the length of the sill. Then, position the new door and door frame on the sill. Make sure the threshold is horizontal. Insert wood shims between the jamb and the wall framing to keep the jamb square.
7. At each shimmed point, drill a counterbored hole through the jamb. Drive a wood screw into each hole and glue the wood plugs in place.
8. Tighten the hinges and make sure the door is still squarely aligned in the jam.
9. Use a utility knife to cut off any excess shim from the edge of the door frame. Next, run a continuous bead of caulk along the gap.
10. Nail the interior trim back around the door and install any exterior trim. Run a continuous bead of caulk along the joint between the sill and the threshold, and between any siding and exterior trim.
11. Install door reinforcers, if desired.
12. Install locks on the door.

Steel doors offer the best protection against fire and break-in attempts. They also offer superior insulation, which helps keep energy costs down. Many steel doors are beautifully designed to look like expensive wood doors. Kalamen doors consist of metal wrapped around wood with a strong frame.

Fiberglass, a strong material that can be made to look like natural wood, offers good resistance to warping and weathering; it's especially useful near pools and saunas or in damp areas. Some types of fiberglass can be stained and finished. Like fiberglass, PVC plastic is strong and isn't affected much by water. However, the plastic surface can be hard to paint.

Aluminum and glass are used together, mostly for sliding glass doors. Although glass can make a door look attractive and it admits light to the interior, it also makes the door less secure.

Among wood doors, the solid-core hardwood types are best. They consist of hardwood blocks laminated together and covered with veneer. A *hollow-core door* provides minimal protection; it consists of two thin panels over cardboard-like honeycomb material. You can recognize a hollow-core door by knocking on it; it sounds hollow. If a burglar kicked a hollow-core door, his foot would go through it.

There's an easy way to reinforce a hollow-core door if aesthetics aren't important. You can clad the exterior side with 12-gauge (or thicker) sheet metal attached with ⁵⁄₁₆-inch-diameter carriage bolts. The bolts should be placed along the entire perimeter of the door about 1 inch in from the door's four edges. Space the bolts about 6 inches apart, and then secure them with nuts on the interior side of the door. If after installing the metal you find the door is too heavy to open and close properly, you may need to remove the hinges and install larger ones.

Another important factor affecting door strength is whether it's flush or paneled. A *flush door* is flat on both sides and is plain-looking. A *paneled door* has surfaces of varying thicknesses, and can be very attractive. The panels may be metal, wood, glass, or a combination of materials. Because the panels are usually thinner and weaker than the rest of the door, they make the door more vulnerable to attack.

You can buy a door, side jambs, trim, threshold, and sill as separate parts together in a single package—a door kit with precut jambs and sills, which is easier to install. You can also buy a door prehung (or preassembled), ready to be fastened to the rough opening.

Many modern door units come with wide light panels (small vertical windows along the sides). If the sidelights might allow a burglar to climb through or to reach in for the lock, they should be made of or lined with a break-resistant material, such as plastic.

Chapter Quiz

1. If a door's hinges can be seen from the exterior side, a burglar might be able to remove them and open the door without touching the lock.

    A. True   **B.** False

2. A flush door is flat on both sides and is plain-looking.

    A. True   **B.** False

3. A hollow-core door provides minimal protection.

    A. True   **B.** False

4. Fiberglass is a strong material that can be made to look like natural wood, and offers good resistance to warping and weathering.

    A. True   **B.** False

5. A paneled door has surfaces of varying thicknesses, and it can be very attractive.

    A. True   **B.** False

6. Aluminum and glass are often used together on sliding glass doors.

    A. True   **B.** False

7. Steel doors offer little protection against fire and break-in attempts, compared to wood doors.

    A. True   **B.** False

8. Glass panels are an especially poor feature in garage doors.

    A. True   **B.** False

9. What kind of door consists of metal wrapped around wood with a strong frame?

    _____

10. A popular reinforcer for door frames is the high-security strike box, a heavy-gauge steel box with 1-inch screws that protrude through the door jamb.

    A. True   **B.** False

11. Typically, an exterior door is connected to its frame by wood hinges on one side and a lock on the other.

    A. True   **B.** False

12. The most secure main garage doors are made of steel, require an automatic door opener, and have no glass or thin panels.

    A. True   **B.** False

13. One type of door reinforcer is a U-shaped metal unit designed to wrap around the door edge under the lock.

    A. True   **B.** False

14. Sliding glass doors are especially vulnerable because their frames and locks are weak.

    A. True   **B.** False

15. Among wood doors, the solid-core hardwood types provide the best security.

    A. True   **B.** False

16. One of the most common ways burglars get through doors is by kicking them.

    A. True   **B.** False

17. In most homes, the style of door depends on the structure's architecture.

    A. True   **B.** False

18. Typically, an exterior door frame consists of a head jamb, two side jambs, stop molding, and a sill or threshold.

    A. True   **B.** False

19. In addition to securing the main door of a garage, any door that allows passage from the garage to the home should be reinforced.

    A. True   **B.** False

20. A hollow-core door provides minimal protection.

    A. True   **B.** False

Chapter 4

SECURING WINDOWS

People who do a lot to secure their doors may be paying little attention to their windows because they think securing windows is time-consuming, expensive, or impossible. To burglars, windows are often the most attractive entry points.

The materials used in making doors are also used for manufacturing window frames. Wood, aluminum, fiberglass, and polyvinyl chloride (PVC) plastic are the most popular for windows. As long as the windows are well built and have good locking devices (keyless types are best), the frame material usually has little effect on a home's security.

Contrary to popular opinion, it usually isn't necessary to make your window frames and panes unbreakable to keep burglars out—unless your neighbors are out of earshot. Burglars know that few things attract more attention than the sound of breaking glass, and they don't like to climb through openings that have large jagged shards of glass pointing at them. When they can't get into a house without breaking a window, most burglars will move on to another house.

You can make your windows more secure just by making them hard to open quietly from the outside. Don't install a lock or any other device that might delay a quick exit in case of a fire. Balancing the safety and security elements depends on what type of windows you have. The four basic types of windows are: sliding, casement, louvered, and double hung.

A *sliding window* works much like a sliding glass door and, like a sliding glass door, it usually comes with a weak lock that's easy to defeat. Most of the supplemental locking devices available for sliding windows fit along the track rail and are secured with a thumbscrew. You can then keep the window in a closed or a ventilating position, depending on where you place the thumbscrew. The need to twist a thumbscrew can be inconvenient if you must frequently lock and unlock a window.

A *casement window* is hinged on one side and swings outward (much like doors do). It uses a crank or a handle for opening and closing. To prevent someone from breaking the glass and turning the crank, the handle should be removed when it isn't being used.

How to Secure Double-Hung Windows at No Cost

1. From inside the home, close the window and clamp the butterfly twist-turn sash lock into the closed position.
2. Use a pencil to mark two spots below the twist-turn lock on the top rail (horizontal member) of the bottom sash. One mark should be about 1 inch inside the left stile; the other should be about 1 inch inside the right stile.
3. Position your drill at the first mark and drill at a slightly downward angle until your drill bit goes completely through the top rail of the bottom sash and about halfway through the bottom rail of the top sash. Then, do the same thing at the other mark you made.
4. Raise the bottom sash about 5 inches and hold it steady. Insert the drill bit back into one of the bottom sash holes and drill another hole about halfway through the top sash. (The hole should be about 5 inches above the other hole you drilled on the stile.) Without moving the sash, do the same thing at the other side of the window.
5. Close the window and insert two small nails or eye bolts into the lower sets of holes to hold the sashes together, so the window can't be lifted open from outside. When you want ventilation, you can remove the nails or bolts, raise the window, and insert them in the top set of holes to secure the window in the open position.

Louvered (or jalousie) windows are the most vulnerable type of window. *Jalousie windows* are made of a ladder-like configuration of narrow, overlapping slats of glass that can easily be pulled out of the thin metal channels. Jalousies attract the attention of burglars and should be replaced with another type of window.

The type of window used in most homes is the *double-hung window*, which consists of two square or rectangular sashes that slide up and down, and are secured with a metal thumb-turn butterfly sash "lock" (although most manufacturers call it a lock, the device is really a clamp). The device holds the sashes together in the closed position, but a burglar can work it open by shoving a knife in the crack between the frames.

Several companies make a useful replacement for conventional sash locks. The devices can't be opened from outside a building. They look like a standard sash lock, but they incorporate a spring-loaded lever that prevents them from being manipulated out of the lock position by using a knife between the sashes.

As an alternative to replacing sash locks, a ventilating wood window lock can be installed. This device lets someone inside raise the window a few inches, and then sets the bolt, which prevents anyone from outside from raising it higher. A *ventilating wood window lock* consists of an L-shaped metal bolt assembly and a small metal base. The bolt assembly fits along the inner edge of either of the two stiles (vertical members) of the top sash and is held in place with two small screws.

The bolt assembly has a horizontal channel that lets you slide the bolt into the locked and unlocked position. When in the locked position, the bolt is parallel to the window and out of the way of the bottom sash. The higher you place the bolt mechanism above the bottom sash, the higher you'll be able to raise the window with the bolt in the locked position. The base isn't really needed, but it helps to prevent the bottom sash from getting marred.

Many companies make ventilating wood window locks, and there aren't important differences between brands. Most models are sold at locksmith shops and home improvement centers for less than $10 each.

Glazing

Glazing is a term that refers to any transparent or translucent material—usually some kind of glass or plastic—used on windows or doors to let in light. Most types of windows can be made more secure by replacing the glazing with more break-resistant material.

The most common glazing for small windows is standard sheet glass. Plate glass—which is a little stronger—is generally used in large picture windows. Because plate glass also has the problem of breaking into many dangerous piece, it shouldn't be used in exterior doors.

Tempered glass is several times stronger than plate glass and costs about twice as much. Rather than shattering into many pieces, tempered glass breaks into small harmless pieces, which is the reason for using it in patio doors. When a large piece of tempered glass breaks, it makes a lot of noise, and this may attract the attention of neighbors.

The strongest type of glass a homeowner might use is laminated glass. *Laminated glass* is made of two or more sheets of glass with a plastic inner layer sandwiched between them. The more layers of glass and plastic, the stronger (and the more costly) the laminated glass is. Laminated glass 4 inches thick can stop bullets and is often used for commercial applications.

Plastics are commonly used as glazing materials. Acrylics, such as Plexiglas and Lucite, are very popular because they are clearer and stronger than sheet glass. However, they scratch easily and can be sawed through.

The strongest types of plastic a homeowner might use are polycarbonates, such as Lexigard and Lexan. Although they're not as clear as acrylics, polycarbonates are up to 30 times stronger. Untreated polycarbonates scratch easily, but you can buy sheets with scratch-resistant coatings.

How to Replace a Glass Pane

1. Working from the exterior side of the window, use a wood chisel to remove any soft or crumbling glazing compound from along the channel between the glass and the window pane. Soften any remaining unpointed glazing compound by applying a heavy coat of linseed oil and waiting about 30 minutes. Then, remove all the unpointed glazing compound. Soften any remaining painted glazing compound by using a heat gun, and then remove the rest of the glazing compound.
2. After all the glazing compound is removed, use a putty knife or long-nose pliers to work the *glazier's points* (small clips used to secure the pane) away from the window pane. Then, lift the pane of glass out of the window.
3. Use a wire brush and medium sandpaper to clean and smooth out each channel. (You may then want to paint the channels for aesthetic reasons.)
4. Insert your new pane of glass or plastic. Seal the pane with whatever type of glazing compound the manufacturer recommends. Peel the paper from the new pane and clean the pane with warm water and a mild detergent.

For increased strength and energy efficiency, many modern windows come in parallel double- or triple-pane configurations. Three parallel panes can provide good security.

Whether you want to replace your glass panes with stronger glass or replace broken panes, you can easily do it yourself. In addition to being unsightly, a broken window can attract burglars because they can quietly remove the glass to gain entry. A broken pane should be replaced immediately.

Glass Blocks

Glass blocks come in a wide variety of patterns and sizes, and they are strong enough to be used in place of plate glass. They're especially useful for securing basement windows. Most patterns create a distorted image to anyone trying to see through them, but clear glass blocks are also available.

For areas that require ventilation, you can buy preassembled panes of glass bock with built-in openings. Preassembled panels are easy to install if you get the right sizes. To order the right size panel, you need to know the size of your window's rough opening. If you have a wood-frame wall, you can determine the rough opening by measuring the width of the opening between the frame's sides, and the height between the sill and the header. You need a panel about ½ inch smaller than that measurement, to make sure it will fit in easily.

If you have a masonry wall, you can determine the rough opening by measuring the width between the brick or block sides and the height between the header and the sill. Be sure the panel you order is about ½ inch smaller than the opening.

You can install glass block panels in two basic ways. The older way involves using masonry cement or mortar—in much the same way as when installing bricks. The newer way involves using plastic strips.

A Newer Way to Install Glass Blocks

A cleaner and simpler way to install glass blocks was recently developed by Pittsburgh Corning. However, the company's glass-block panel kits aren't designed to offer strong resistance to break-in attempts. The kit includes: U-shaped strips of plastic channel for the perimeter of the panel, a roll of clear plastic spacer for holding the blocks in place, clear silicon caulk, and a joint-cleaning tool. The kit can be used in the following way:

1. Install the U-shaped channels, using shims if necessary. Attach the channels with 1-inch flat-head wood screws, and conceal the screw heads with white paint.

2. Using short lengths of spacer, align the blocks vertically. (Use a utility knife to cut spacers to size.)

3. Align the blocks horizontally with long strips of spacer. (The spacer's contour matches the block edge surface.)

4. Fit the last block into the panel through a section temporarily removed from the top channel.

5. Slip the leftover section of the top channel over the last block, and apply caulk to hold the section in place.

6. Wipe the joints clean with a cloth dampened with isopropyl alcohol. Fill the joints with beads of silicon caulk.

How to Install Glass Blocks with Mortar

1. For a masonry wall installation, insert shims in the opening around the glass block panel. Make sure the panel is squarely in place.
2. Use a trowel to push mason's mortar into the gaps around the perimeter of the panel.
3. After the mortar sets, remove the shims and fill their spaces with mortar. Then, seal the panel with a continuous bead of caulking.

Protecting Glass

If you're concerned that someone will gain entry by breaking a window, you might want to coat the window with security film, a transparent laminated coating that resists penetration and firmly holds broken glass in place. Even after breaking the glass, burglars would have a hard time getting through. Some security film can hold glass in place against sledgehammer blows, high winds, and explosions.

A less-aesthetic, but equally effective way to protect windows is to install iron security bars. The mounting bolts should be reachable only from inside your home. Be sure the bars don't make it hard for you to escape quickly, if necessary. Hinge kits are available for many window bars.

Before buying window bars, you need to measure the width and height of the area to be covered. In general, the larger the area to be protected, the more the bars will cost. Window bars range in price from about $10 to more than $50 per window.

Chapter Quiz

1. Which type of glass is the strongest?

    A. Laminated glass

    B. Sheet glass

    C. Tempered glass

2. Which type of window is the least secure?

    A. Double-hung window with a wood ventilating lock

    B. Louvered window

    C. Casement window with the handle removed

3. The term "glazing" refers to any transparent or translucent material.

    A. True   **B.** False

4. Tempered glass is weaker than plate glass.

    A. True   **B.** False

5. What is the most common type of window used in homes?

    _____

6. Which window is made of a ladder-like configuration of narrow, overlapping slats of glass?

    _____

7. Which window is hinged on one side and swings outward (much like doors do)?

    _____

8. Two strong types of polycarbonates are Lexigard and Lexan.

    A. True   **B.** False

9. Laminated glass is made of two or more sheets of glass with a plastic inner layer sandwiched between them.

    A. True   **B.** False

10. Acrylics, such as Plexiglas and Lucite, are clearer and stronger than sheet glass.

    A. True   **B.** False

11. The most common glazing for small windows is standard sheet glass.

    A. True   **B.** False

12. Glass block is especially useful for securing basement windows.

    A. True   **B.** False

13. For areas that require ventilation, you can buy preassembled panels of glass block with built-in openings.

    A. True   **B.** False

14. Security film is a laminated coating that resists penetration and firmly holds broken glass in place.

    A. True   **B.** False

15. What does a ventilating wood window lock do?

    _____

16. It's usually important to make every window in a home unbreakable to keep intruders out.

    A. True   **B.** False

Chapter 5
BASIC LOCKS
AND KEYS

Terms such as "mortise bit-key lock" and "Medeco key-in-knob lock" mean little to most people, but they provide useful information to locksmiths. Like other trades, locksmithing has its own vocabulary to meet its special needs.

Terminology

Laypersons frequently use a generic name, such as padlock, automobile lock, or cabinet lock, when referring to a lock. Such a name has limited value to locksmiths because it is so general. It simply refers to a broad category of locks that are used for a similar purpose, share a similar feature, or look similar to one another.

Generic Names

Some of the most commonly used generic lock names include automobile lock, bike lock, ski lock, cabinet lock, deadbolt lock, gun lock, key-in-knob lock, luggage lock, lever lock, padlock, combination lock, and patio door lock. Sometimes generic terms have overlapping meanings. A padlock, for instance, can also be a combination lock.

The *key-in-knob* lock refers to a style of lock operated by inserting a key into its knob. A *lever lock* has a lever as a handle. A handleset has a built-in grip handle. A *deadbolt lock* projects a deadbolt. As the names imply, the automobile lock, bike lock, ski lock, patio door lock, and so forth are based on the purposes for which the locks are used. Sometimes, locks that share a common purpose look very different from one another.

Manufacturers' Names

Locksmiths often refer to a lock by the name of its manufacturer, especially when all or most of the company's locks share a common characteristic. Locks manufactured by Medeco Security Locks, Inc., for example, all have similar internal constructions. Simply by knowing a lock is a Medeco lock, a locksmith can consider the options for servicing it.

Several lock manufacturers are so popular in the locksmithing industry that every locksmith is expected to be familiar with their names and the common characteristics of each manufacturer's locks. Those manufacturers include Arrow, Best, Corbin, Dexter, Ilco Unican, Kwikset, Master, Medeco, Russwin, Sargent, Schlage, Weiser, and Yale.

Type of Key

Many times, a lock is identified by the type of key used to operate it. Bit key locks and tubular key locks are two common examples. Tubular key locks, sometimes called *Ace locks*, are primarily used on vending machines and coin-operated washing machines. Bit key locks are used on many closet and bedroom doors. When speaking about a bit key lock, locksmiths usually use a name that reveals how it is installed.

Installation Method

The terms "rim lock" and "mortise lock" identify locks based on their installation method. A *rim lock*, or surface-mounted lock, is designed to be installed on the surface, or rim, of a door.

A *mortise lock* is designed to be installed in a mortise, or recess, in a door. Not all mortise locks are operated with a bit key.

Internal Construction

For servicing locks, names based on their internal constructions are usually most helpful to a locksmith. Examples include warded lock, pin tumbler lock, disc tumbler lock, wafer tumbler lock, lever tumbler lock, and side bar lock.

Lock names based solely on internal construction don't indicate the lock's purpose, installation method, function, or appearance. They only refer to its type of cylinder or parts inside its lock case. A lock that uses a pin tumbler cylinder, for example, is called a *pin tumbler lock* or a *pin tumbler cylinder lock*. A lock with wards inside its case is called a *warded lock*.

Note: Some people use the terms "lever lock" and "lever tumbler lock" synonymously. However, *lever lock* refers to a type of handle used, whereas *lever tumbler lock* refers to a type of internal construction.

Most types of cylinders can be used with a wide variety of locks. A key-in-knob lock, for example, can use a disc tumbler cylinder or a pin tumbler cylinder. Both cylinder types can also be used with many other types of locks. Which type of cylinder is best to use depends on the level of security needed, how much money someone is willing to spend, and whether or not the cylinder needs to fit into an existing keying system.

Pin Tumbler Locks

A *pin tumbler lock* is any lock that relies on the pin tumbler cylinder as its primary means of security. The pin tumbler lock comes in a wide variety of shapes and is used for many purposes—such as building doors, automobile doors, and ignitions. Key-in-knob locks, deadbolt locks, and padlocks are often pin tumbler locks. To identify a pin tumbler lock, you can look through the keyway. Typically, you can see the first pin.

Parts of a pin tumbler cylinder include: the cylinder case (or "housing" or "shell"), plug (or "core"), keyway, lower-pin chambers, springs, top pins (or "drivers"), and bottom pins. Some pin tumbler cylinders have more parts, but all models work basically the same way.

The cylinder case houses the other basic parts. The *plug* is the part with the keyway, which rotates when you're turning the key. When the cylinder is disassembled, you can see drilled holes, usually five or six, along the length of the plug. Those are *lower-pin chambers* and each holds a tapered bottom pin. Inside the cylinder case, in alignment with the lower-pin chambers, are holes that correspond in size and position, called upper-pin chambers. The *upper-pin chambers* each house a spring that presses against one or more top pins. Each set of top and bottom pins within corresponding pin chambers is called a stack. Usually each stack has only two pins (a top and a bottom), but master-keyed cylinders may have three or more pins in a stack.

The position of each pin determines whether or not the cylinder can be rotated, which is necessary to unlock the lock. Pin positions are determined by gravity, pressure from the springs, and pressure from the key (or lock pick). When no key is in the keyway, gravity and the downward pressure of the springs drive the top pins into the plug, until they rest on their corresponding bottom pins. Because the bottom pin lengths vary from one lower-pin chamber to another, some top pins will drop different depths into the lower-pin chambers. When a pin is in its upper-pin chamber and in a lower-pin chamber at the same time, the pin obstructs the plug from turning. If you tried to rotate the plug forcibly at that time, you would likely bend the pin, causing a serious lockout.

When a cylinder is made, space always exists between the case and its plug. Otherwise, the plug would be jammed in so tight, it could never be turned, regardless of the pin positions or which key is used. That space between the case and the plug is called a *shear line*. When inserted into the keyway, the proper key slides under all the bottom pins and lifts each to the shear line. The proper key will fit the keyway and have properly spaced cuts of the right depths to match each bottom pin length. When all the top and bottom pins meet at the shear line, none of them are obstructing the plug from being rotated. When the plug is rotated, the top pins separate from their respective bottom pins.

Lock Functions

Entrance lock, classroom lock, and vestibule lock are names based on how a lock functions. A classroom lock, for instance, is one whose inside knob is always in the unlocked position for easy exiting and whose outside knob can be locked or unlocked with a key. An institutional lock, however, has both knobs always in the locked position to prevent easy exiting; a key must be used on either knob to operate the lock.

Naming Conventions

At this point, you should have a good idea of how locksmiths identify locks. They simply combine several applicable terms that provide the necessary specificity.

Locksmiths identify a lock in ways that convey information needed to purchase, install, and service it. The name they use is based not only on the purpose and appearance of the lock, but also on the lock's manufacturer, key type, method of installation, type of internal construction, and function.

The names used by a locksmith are typically formed by combining several words. Each word in the name provides important information about a lock. The number of words a locksmith uses for a name depends on how much information they need to convey.

When ordering a lock, for instance, the locksmith needs to use a name that identifies the lock's purpose, manufacturer, key type, appearance, and so forth. However, a name that simply identifies the lock's internal construction may be adequate for describing a servicing technique to another locksmith.

Types of Keys

The most common keys, those used on homes and businesses, share many common features. Usually, such keys are made of metal, are 2 to 3 inches long, and have the following parts: a *bow* (the gripping part for turning the key), a thin blade with grooves or "millings" on one or both sides, and jagged U- or V-shaped cuts of varying depths spaced along one or both edges of the blade. If you look at your keys, you'll probably find most fit that description.

Keys come in many other shapes and sizes for operating a wide variety of locks. Some keys for low-cost magnetic padlocks, for instance, are thin rectangular bars about 2 inches long. Keys for some electric locks are roughly the size and shape of a dime. Electronic door locks at many hotels use thin plastic keys, roughly the size and shape of a playing card.

A locksmith doesn't need to know everything about all the different kinds of keys, but becoming familiar with the basic types is a good idea. Locksmiths commonly sell and work with eight basic types of keys: bit key, barrel key, flat key, corrugated key, cylinder key, tubular key, angularly bitted key, and dimple key. Virtually all other mechanical keys are variations of these types.

Bit Keys

A *bit key* is used for operating bit key locks. It is usually made of iron, brass, or aluminum. This key is sometimes called a *skeleton key*. The main parts of a bit key are the bow, shank, shoulder, throat, post, and bit, tumbler cut and ward cuts.

Barrel Keys

Many barrel keys look similar to bit keys. They have many of the same parts. The major difference between the two types of keys is that the *barrel key* has a hollow shank. Another difference is that barrel keys don't have a shoulder, post, or blade.

Flat Keys

As the name implies, a *flat key*, or flat steel key, is flat on both sides. Most are made of steel or nickel silver. Such keys are often used for operating a lever tumbler lock, a type of lock used on luggage and safe deposit boxes.

Corrugated Keys

Many corrugated keys look similar to flat keys. Both types usually have the same parts. But *corrugated keys* have corrugations, or ripples, along the length of their blades. They are designed to let the key fit into correspondingly shaped keyways. Unlike most flat keys, corrugated keys have cuts on both sides of their blades.

Corrugated keys are often used with warded padlocks. Some corrugated keys are designed to operate other types of locks. For example, Schlage Lock Company used to manufacture a key-in-knob lock that used special types of corrugated keys. Those keys look more like cylinder keys than flat keys.

Cylinder Keys

The most popular key today is the *cylinder key*, which is used to operate pin tumbler locks and disc tumbler locks. You probably have several cylinder keys to unlock the front door of your home or the doors of your car.

The parts of a cylinder key are the bow, shoulder, blade, tumbler cuts, keyway grooves, and tip. The *shoulder* acts as a stop; it determines how far the key will enter the keyway. Some cylinder keys don't have shoulders; those keys use the tip as a stop. The *keyway grooves* are millings along the length of a key blade that allow the key to enter a keyway of a corresponding shape.

Tubular Keys

The *tubular key* has a tubular blade with cuts, or depressions, milled in a circle around the end of the blade. The key is used to operate tubular key locks, which are often found on vending machines and coin-operated washing machines.

Tubular keys are often improperly called Ace keys. The term "Ace key" is short for a brand name, but it doesn't apply to all tubular keys. The first tubular key was patented by the Chicago Lock Company to operate its Chicago Ace Lock brand tubular key lock. Today, many companies manufacturer tubular key locks and tubular keys.

Parts of a tubular key include the bow, blade, tumbler cuts, and nib. The *nib* shows which position the key must enter the lock to operate it. The purposes of the bow, blade, and tumbler cuts are similar to the purposes of corresponding parts of a cylinder key.

Angularly Bitted Keys

The *angularly bitted key* is used with some high-security locks. The key has cuts that angle perpendicularly from the blade. The key is designed to cause tumblers within a cylinder to rotate to specific positions. Medeco Security Locks, Inc. popularized the angularly bitted key.

Dimple Keys

The *dimple key* is used to operate some high-security pin tumbler locks. It has cuts that are drilled or milled into its blade surface; the cuts normally don't change the blade's silhouette. Lori Corporation's Kaba locks are popular locks operated with dimple keys.

Chapter Quiz

1. Parts of a tubular key include the bow, blade, tumbler cuts, and nib.

   A. True   **B.** False

2. The parts of a cylinder key are the bow, shoulder, blade, tumbler cuts, keyway grooves, and tip.

   A. True   **B.** False

3. The most popular key today is the bit key.

   A. True   **B.** False

4. A flat key, or flat steel key, is flat on both sides.

   A. True   **B.** False

5. The dimple key is used to operate some high-security pin tumbler locks.

   A. True   **B.** False

6. Medeco uses angularly bitted keys.

   A. True   **B.** False

7. The tubular key has a tubular blade with cuts, or depressions, milled in a circle around the end of the blade.

   A. True   **B.** False

8. Many barrel keys look similar to cylinder keys.

   A. True   **B.** False

9. Bit keys are often called skeleton keys.

   A. True   **B.** False

10. Parts of a pin tumbler cylinder include: the cylinder case (or "housing" or "shell"), plug (or "core"), keyway, lower-pin chambers, springs, top pins (or "drivers"), and bottom pins.

    A. True   **B.** False

11. Lock names, such as padlock, automobile lock, and cabinet lock, are generic names for locks.

   A. True   **B.** False

12. The key-in-knob lock refers to a style of lock operated by inserting a key into its knob.

   A. True   **B.** False

13. A lever lock has a lever as a handle.

   A. True   **B.** False

14. A deadbolt lock projects a spring bolt.

   A. True   **B.** False

15. Bit key locks and tubular key locks are names based on the type of key used to operate the lock.

   A. True   **B.** False

Chapter 6
PICKING, IMPRESSIONING, AND BUMPING LOCKS

T his chapter explains how to pick open, impression, and bump open standard pin tumbler locks.

Picking Pin Tumbler Locks

With practice, you should be able to pick open most standard pin tumbler locks within a few minutes. In theory, any mechanical lock operated with a key can be picked because tools and techniques can be fashioned to simulate the action of a key.

A simple way to describe lock picking is to insert a lock pick and a torque wrench into a lock plug in such a way that the pick lifts each tumbler into place (where the right key would place them), while the torque wrench provides the pressure to turn the plug into the open position. You have to vary the pressure of the torque wrench. If you turn it too hard, the pins won't move into position. If you turn it too lightly, the pins won't stay in place. Usually, the problem is the torque wrench is turned too hard.

Lock picks come in a wide variety of shapes and sizes. Some are for specific locks, such as tubular key locks. The key to picking locks fast is to focus on what you're doing and to visualize what's happening in the lock while you're picking it. If you know how pin tumbler locks work, it's easy to understand the theory behind them.

A common method of picking locks is the rake method, or raking. To *rake* a lock, insert a pick (usually a half diamond or rake) into the keyway past the last set of pin tumblers, and then quickly move the pick in and out of the keyway in a figure-eight movement, while varying tension on the torque wrench. The scrubbing action of the pick causes the pins to jump up to (or above) the shear line, and the varying pressure on the torque wrench helps catch and bind the top pins above the shear line. Although raking is based primarily on luck, it sometimes works well. Many times, locksmiths rake a lock first, to bind a few top pins, and then pick the rest of the pins.

Using a Lock Pick Gun

A *pick gun* can be a great aid in lock picking. To use a pick gun, insert its blade into the keyway below the last bottom pin. Hold the pick gun straight, and then insert a torque wrench into the keyway. When you squeeze the trigger of the pick gun, the blade slaps the bottom pins, which knocks the top pins into the upper pin chambers. Immediately after each squeeze, vary the pressure on the torque wrench. You will likely capture one or more upper pins in their upper-pin chambers and set them on the plug's ledge. Then, you can pick each of the remaining pin sets, one by one.

Before attempting to pick a lock, make sure the lock is in good condition. Turn a half-diamond pick on its back, and then try to raise all the pin stacks together. Then, slowly pull the pick out to see if all the pins drop, or if one or more of the pins are frozen. If the pins don't all drop, you may need to lubricate the cylinder or remove foreign matter from it.

Hold the pick as you would hold a pencil—with the pick's tip pointing toward the pins. With the other hand, place the small bent end of a torque wrench into the top or bottom of the keyway, whichever position gives you the most room to maneuver the pick properly. Make sure the torque wrench doesn't touch any of the pins. Use your thumb or index finger of the hand that's holding the torque wrench to apply light pressure on the end of the torque wrench in the direction you want the plug to turn.

While using a pick, carefully lift the last set of pins to the shear line, while applying slight pressure with the torque wrench. The *shear line* is the space between the upper- and lower-pin chambers. Take a mental note of how much resistance you encountered while lifting the pin stack. Release the torque wrench pressure, letting the pin stack drop back into place. Then, move on to the next pin stack and do the same thing, keeping in mind which pin stack offered the most resistance. Repeat that with each pin stack.

Next, go to the pin stack that offered the most resistance. Lift the top of its bottom pin to the shear line, while varying pressure on the torque wrench. Apply enough pressure on the torque wrench to hold that picked top pin in place. Then, gently move on to the next most-resistant stack. Continue lifting each pin stack (from most resistant to the least resistant) to the shear line. As you lift each pin stack into place, you are creating a larger ledge for other top pins to rest on. When all the top pins are resting on the plug, the plug will be free to turn to the unlocked position.

Hands-On Experience

No amount of reading can make you good at picking locks. You need to practice often, so you develop the sense of feel. You need to learn how to feel the difference between a pin tumbler that has been picked (that is, placed on the ledge of the plug) and one that is bound between its upper and lower chambers.

To practice lock picking, start with a cylinder that has only two pin stacks in it. When you feel comfortable picking that, add another pin stack. Continue adding pin stacks until you can pick at least a five-pin tumbler.

When you're practicing, don't rush. Take your time, and focus on what you're doing. Always visualize the inside of the lock and try to picture what's happening while you're picking the lock. For the best results, practice under realistic circumstances. Instead of sitting in a comfortable living room chair trying to pick a cylinder, practice on locks on a door or on a display mount.

With a lot of focused practice, you'll find yourself picking all kinds of locks faster than ever.

Impressioning Locks

From the outside, *impressioning* is inserting a prepared key blank fully into a keyway, and then twisting the blank clockwise and counterclockwise, in a way that leaves tumbler marks on the blank, which shows where to file the blank to make a working key. You then file the marks, clean the blank, reinsert it, twist it again, and then file it at the new marks. At some point, you'll have a working key. With practice, you should be able to impression most pin tumbler locks within five to ten minutes.

Pin Tumbler Locks

To impression a pin tumbler lock, you first need to choose the right blank (one that fits fully into the keyway). If the blank is too tight, you won't be able to rock it enough to mark it. The blank also needs to be long enough to lift all the pins. If you use a five-pin blank on a six-pin cylinder, you probably won't be able to impression it because the sixth pin won't mark the blank. To choose the right size blank, use a probe or pick to count the number of pin sets in the lock.

The material of the blank needs to be soft enough to be marked by the pins, but not so soft as to break off while you're twisting or rocking the blank. Nickel-silver blanks are too hard for impressioning because they don't mark well. Aluminum blanks are soft enough to mark well, but they break off too easily. Brass blanks work best. Nickel-plated brass blanks are also good for impressioning because the nickel plating can be filed off.

Filing the Blank

New key blanks have a hardened glazed surface that hinders impressioning, unless you prepare them. To prepare a blank, shave the length of the blade along the side that comes in contact with the tumblers. Shave the blank at a 45-degree angle without going too deep into the blank. You want the blank's biting edge to be sharp (a knife edge) without reducing the width of the blank. File forward only. Don't draw the file back and forth across the blank!

Use a round or pippin file with a Swiss No. 2, 3, or 4 cut. A coarser file will shave the blank quicker, but it leaves rougher striations on the blank—making it harder to see the pin marks. A finer file makes the marks easier to see, but it clogs quicker while you're filing. This makes impressioning take a long time. You probably won't find impressioning files at a hardware store or a home improvement center, but they're sold through locksmith-supply houses.

Another popular way to prepare the blank is to turn the blank over and shave the other side along the length where the tumblers touch. After shaving both sides of the biting edge at 45 degrees, you'll have a double-knife edge.

Other Useful Equipment and Supplies

In addition to a file, you need a key-holding device, such as an impressioning tool or a 4- or 5-inch pair of locking pliers. A magnifying glass can be helpful for seeing impressioning marks. A head-wearing type magnifying glass lets you see the marks and file the marks at the same time. Although they aren't essential, you can use depth and space charts, as well as a caliper to file marks more precisely.

Key Bumping

One of the controversial subjects among locksmiths is whether or not to alert the general public about a lock-opening technique called "key bumping." This technique is controversial because, unlike lock picking and key impressioning, *key bumping* is easy to learn and is effective on most pin tumbler locks (including high-security locks).

How to Bump Locks

Bumping locks requires a key that fits all the way into the plug and a tool to tap on the key bow. To prepare the key (or blank), you need to cut each space to its deepest depth. For many keys, that's the "9" depth—which is why some people call bump keys "999" keys. To make a bump key for a lock that uses a "6" as the deepest cut, for example, you cut a "6" at each space.

You don't need a new key blank to make a bump key. You can use any key that goes all the way into the plug and uses standard-cut depths for the lock. The easiest way to make a bump key is to use a code machine or a code key cutter. If you don't have one, ask a local locksmith to make bump keys for you. You can also make them with a file, a caliper, and depth and space information. To get the right depth and spaces for the cuts, you can use depth and space charts. The space charts show you where to place the cuts along the length of the key. The depth charts show you how deep to make the cuts.

When making a bump key, be careful not to cut any space too deep. File a little metal off the shoulder stop of the key (about 0.25 inches) on the key's biting side. Don't cut too much off the key. You can always cut off a little more, but you can't add metal to a key.

The first step to using the bump key is to fully insert it into the lock's plug. Then, pull the key out until you hear or feel one click. Use one of your hands to provide turning pressure on the key bow, while simultaneously tapping the back of the bow with a screwdriver handle, a tool designed to bump keys, or a small hammer. Tap the key hard enough for it to go fully into the plug.

If the plug doesn't turn, remove the key, and then reinsert it all the way. Pull the key out until you hear one click. Then, again apply turning pressure, while tapping the bow. Sometimes, you may need to tap the bow a few times to open a lock.

Most standard locks can be quickly bumped open, but the technique doesn't work on all locks—especially locks that require two or more actions to occur to open the lock, for example, locks that require pins to meet at the shear line, while simultaneously requiring the movement of side pins, side bars, or rotating elements.

Bumping keys works similarly to picking a lock with an electric or manual pick gun. By hitting the key bow, the bottom pin slides up into the top of the cylinder to momentarily create a shear line, thus allowing the cylinder plug to turn if you have the right amount of turning pressure.

Chapter Quiz

1. Most standard pin tumbler locks can be quickly bumped open.

   A. True   **B.** False

2. Given enough time and proper tools, any mechanical lock operated with a key can be picked.

   A. True   **B.** False

3. A common method of picking locks is the rake method.

   A. True   **B.** False

4. To impression a pin tumbler lock, you first need to choose a blank that's too tight to rock it and short enough to lift all but one set of pins.

   A. True   **B.** False

5. The *shear line* is the space between the upper- and lower-pin chambers.

   A. True   **B.** False

6. To choose the right size blank for impressioning a lock, use a probe or pick to count the number of pin sets.

   A. True   **B.** False

7. To impression a lock, the material of the blank needs to be hard enough not to be marked by the pins.

   A. True   **B.** False

8. Nickel-silver blanks are usually best for impressioning.

   A. True   **B.** False

9. To impression a lock, you need a file and a key-holding device, such as an impressioning tool or a pair of locking pliers.

   A. True   **B.** False

10. To prepare a blank for impressioning, you need to shave the length of the blade along the side that comes in contact with the tumblers.

    A. True   **B.** False

11. One of the controversial subjects among locksmiths is whether or not to alert the general public about a lock-opening technique called "key bumping." This is because, unlike lock picking and key impressioning, key bumping is easy to learn and is effective on most pin tumbler locks (including high-security locks).

A. True **B.** False

12. Bumping locks requires a key that fits all the way into the plug and a tool to tap on the key tip.

A. True **B.** False

13. You need a new key blank to make a bump key.

A. True **B.** False

14. You can make bump keys with a file, a caliper, and depth and space charts.

A. True **B.** False

15. Most standard locks can be quickly bumped open.

A. True **B.** False

16. Bumping keys works similarly to picking a lock with a pick gun.

A. True **B.** False

17. Some locksmiths call bump keys "999" keys.

A. Truc **B.** False

18. The best files for impressioning are a round or pippin file with a Swiss No. 2, 3, or 4 cut.

A. True **B.** False

19. When making a bump key, it's important not to cut any space too deep.

A. True **B.** False

20. Although they aren't essential, you can use depth and space charts along with a caliper to file impression marks more precisely.

A. True **B.** False

Chapter 7

ELECTROMAGNETIC LOCKS

The electromagnetic lock was introduced in the United States in 1970 and has gained considerable popularity. Today, it's a popular part of access control systems throughout the world.

Electromagnetic locks are often used to secure emergency exit doors. When connected to a fire alarm system, the lock's power source is automatically disconnected when the fire alarm is activated. This allows the door to open freely, so people can exit quickly.

Although the principle of operation of an electromagnetic lock is different from that of a conventional mechanical lock, the former has proven a cost-effective, high-security locking device. Unlike a mechanical lock, an electromagnetic lock doesn't rely on the release of a bolt or a latch for security. Instead, it relies on electricity and magnetism.

Structure

A standard electromagnetic lock consists of two components: a rectangular electromagnet and a rectangular, ferrous-metal strike plate. The electromagnet is installed on a door's header; the strike plate is installed on the door in a position that allows it to meet the electromagnet when the door closes. When the door is closed and the electromagnet is adequately powered—usually by 12 to 24 direct-current (DC) volts at 3 to 8 watts—the door is secured. Typically, the locks have 300 to 3000 pounds of holding power.

Security Features

One of the biggest fears people have about electromagnetic locks is power failure. What happens if the power goes out or if a burglar cuts the wire connecting the power source to the lock? In such cases, standby batteries are often installed with the lock to provide continued power. Also, the lock can't be tampered with from outside the door because it is installed entirely inside the door. No part of the lock or power supply wires is exposed from outside the door.

Another important security feature of electromagnetic locks is they are fail-safe. That is, when no power is going to the electromagnet, the door will not be locked. That is why the lock meets the safety requirements of many North American building codes.

Disadvantages

Electromagnetic locks have two major disadvantages. First, the locks often cost from four to ten times more than typical high-security mechanical locks. And, second, many people think electromagnetic locks are much less attractive than mechanical locks.

When using electromagnetic locks in a typical application, additional pieces of hardware must be installed to comply with building fire codes. These fire codes are used throughout the United States. The code states the following: There must be a minimum of two devices used to release the electromagnetic lock. One device must be a manual release button that has the words "PUSH TO EXIT" labeled.

This push button must provide a 30-second time delay when pushed, and the time delay must act independently of the access control system (the delay must work on its own and not be tied into any other access control system). Another device can be either a PIR motion detector or an electrified exit release bar (also called a *crash bar*). If the building has a fire alarm system, the electromagnetic lock must be tied into the fire-control system, so the lock unlocks automatically during a fire alarm. The door must also release on the loss of main electrical power.

The DS-1200 Electromagnetic Lock

The DS-1200 model, made by Highpower Security Products, has 1200 pounds of holding power, and is for external or industrial applications. It can be used in harsh environments to secure doors and gates. All electronics are sealed in epoxy and are protected by the steel housing cover. The housing armature and exposed face of the electric lock are nickel plated to resist rust and corrosion. A rigid conduit fitting can be provided on one end of the lock to protect power wiring in gate-control installations.

All DS-1200 electromagnetic locks are fail-safe, releasing instantly on command or loss of power. There are no moving parts to wear, stick, or bind.

The rugged design and durable construction of this lock assures virtually unlimited actuations without fear of electrical fatigue or mechanical breakdown. Proudly made in the U.S.A., a ten-year limited warranty is provided.

The standard *DS-1200 model* is supplied with an adjustable mounting plate for use on out-swinging doors. The *DS-1200-TJ unit* is furnished with an angle lock mounting plate and an armature Z bracket for in-swinging door installations. Any 1200 series lock may be converted in the field for in-swinging door applications by adding the angle mounting plate and Z bracket. An optional conduit-fitted lock is available for exterior gate-control applications.

All 1200 series locks can be operated on either 12 or 24 VDC. The efficient design of these locks requires only 170 milliamp (ma) at 24 volts DC to maintain the rated 1200-lb. holding force.

Because they can be controlled individually or simultaneously from one or several locations, these locks are ideal for securing manual or automatically operated doors and gates.

The Thunderbolt 1500

The Thunderbolt 1500, made by Highpower Security Products, is a 2-inch profile electromagnetic lock for out-swinging doors. Designed to provide fast installation, the unit incorporates a tamper-resistant cover design and slotted mounting system with installation template. Providing 1500 pounds of holding force, all versions feature a replaceable dual-voltage coil that operates at both 12 and 24 volts. Units can be equipped to operate using either alternating current (AC) or DC power, and they feature a surge- and spike-suppression circuit. The Thunderbolt 1500 is made in the U.S. and is backed with a ten-year manufacturer's warranty.

The lock is available in over 150 colors of powder-coat finishes and in solid brass by special order. This is a slim-line 1500-pound electromagnetic lock, designed to provide the fastest installation times. Mounting the Thunderbolt is quick and easy. Extensive feedback from installers directed Highpower to incorporate an improved adjustable mounting system. This slotted mounting system allows freedom of movement to make adjustments during tough installations. A template is provided to quickly mark mounting holes for both the magnet and the armature.

Installers are discovering that the Thunderbolt 1500 is designed to maximize profits. The Thunderbolt 1500 has an epoxy-less design that allows the magnetic coil to be unplugged from the unit and replaced if it becomes defective, without having to uninstall the lock. This feature both reduces service time and provides improved value, by keeping assembly costs low.

No more fooling around with wire splices. Connections are made to the Thunderbolt with screw terminal blocks that speed wire installation. With a single circuit board, the Thunderbolt can quickly be configured with a door position switch (DPS), a cover tamper switch (CTS), and a magnetic bond sensor (MBS). In addition, electronic spike and "kickback" surge suppression is standard with all models.

Installers love the Thunderbolt's single-piece cover. It slides into place, allowing a rapid and hassle-free installation. Because it has no exposed screws, the cover provides the highest level of tamper resistance. This modular cover allows installers to stock different color covers to quickly provide customers with the cosmetics specified.

Chapter Quiz

1. The two major disadvantages of electromagnetic locks are

    _____.

2. Unlike a mechanical lock, an electromagnetic lock doesn't rely on the release of a bolt or a latch for security.

    A. True   **B.** False

3. Connections are made to the Thunderbolt with screw terminal blocks that speed wire installation.

    A. True   **B.** False

4. Thunderbolt's single-piece cover slides into place, allowing a rapid and hassle-free installation.

    A. True   **B.** False

5. The Thunderbolt 1500 has an epoxy-less design that allows the magnetic coil to be unplugged from the unit and replaced if it becomes defective, without having to uninstall the lock.

    A. True   **B.** False

6. The Thunderbolt is available in over 350 colors of powder-coat finishes and in solid brass by special order.

    A. True   **B.** False

7. The Thunderbolt 1500 is designed to provide the fastest installation. The unit incorporates a unique tamper-resistant cover design and slotted mounting system with installation template.

    A. True   **B.** False

8. All Thunderbolt 1200 series locks can be operate on either 12 or 24 VDC.

    A. True   **B.** False

9. A template is provided with the Thunderbolt electromagnet locks to quickly mark mounting holes for both the magnet and the armature.

    A. True   **B.** False

10. Electronic spike and "kickback" surge suppression is standard with all Thunderbolt models.

    A. True   **B.** False

Chapter 8

OPENING AUTOMOBILE DOORS

While the automobile has been with us since the beginning of the twentieth century, the lock was adopted slowly. However, by the late 1920s, nearly every auto had an ignition lock, and closed cars had door locks as well. Current models can be secured with half a dozen locks. This chapter explains how to open and service all kinds of vehicles.

Opening Locked Cars

Car opening can be a lucrative part of any locksmithing business. For some, it's the biggest source of income. To offer car-opening services, you only need a few inexpensive tools and some technical knowledge. In this chapter, I show you how to buy and make the tools you need, and I give you detailed instructions on how to open most cars. I also tell you about the business matters you need to know.

In the interest of self-disclosure, I should point out that, several years ago, I was hired by a major automotive lock manufacturer to prepare and edit its car-entry manual, which included creating new entry techniques and designing tools. At the time, it was one of the most comprehensive and best-selling publications of its kind. Although the manual is out-of-date, copies are still being sold. I no longer work on that publication. The suggestions and tool designs I give here are original and they aren't meant to promote any company's products.

Tools You Need

Car-opening tool sets sold through locksmithing supply houses may include 40 or more tools. Toolmakers point out that the variety is necessary (or at least helpful) because of the constant lock-related changes made to new cars. Some of the uniquely shaped tools are designed for one specific make, model, and/or year of car. Whether or not all the new specialized tools are worth the money is debatable. But, a continuous supply of new tools means recurring revenue for the toolmakers.

You can open most cars with only five simple tools, all of which you can make yourself. In some cases, not only is this cheaper, but the tool will work better if you make it yourself. Later, I tell you how to make the tools. The most important car-opening tools are a slim jim, which is a hooked horizontal linkage tool, an *L* tool, a *J* tool, and an across-the-car tool (aka a long-reach tool). They have different ways of reaching and manipulating a car's lock assembly.

Slim Jim

The *slim jim* is a flat piece of steel with cutouts near the bottom on both sides. The cutouts let you hook and bind a linkage rod from either side of the tool. The tool can also be used to push down on a lock pawl. Slim jims come in different widths, and it's good to have both a wide one and a thin one. You can buy slim jims at most automobile supply and hardware stores, but you can get better models from a locksmith supply house. They're often sturdier, have more notches, have a handle, and just generally look more professional. To make your own, you need a 24-inch piece of flat steel or aluminum, from 1 to 2 inches wide. You can use a ruler or another item that's the right size made from the proper material. Just draw the slim-jim shape onto the metal, and then grind away the excess material.

Hooked Horizontal Linkage Tools

Hooked horizontal linkage tools go by many names and come in all kinds of sizes and configurations: the small hook on the end of the tool lets you catch and bind a horizontal rod and slide it to unlock the door. Some hook down onto the rod; others hook from the bottom of the rod. It's good to have both of them. Two other kinds of horizontal linkage tools include the three "fingers" type that spreads to clamp onto the rod, and the tooth-edged type that bites into the rod. I don't like either of those two because when using them, you have to be especially careful to avoid bending the linkage rods.

J Tool

The *J tool* is one of the easiest to use: it goes within the door, between the window and weather stripping. Then, it goes under the window and beneath the lock button to push the button up to the unlocked position.

L Tool

An *L* tool is used to push or pull on bell cranks and lock pawls. You can also use it to access the lock rod by going under the lock handle.

> **Note:** For versatility, buy or make a tool that is an *L* tool on one end and a *J* tool on the other. The part of the tool that enters between the door needs to be a specific shape; the rest of the tool is the handle. Making or buying tools that have a different or a different-size tool on each end is useful.

Across-the-Car Tool

The across-the-car tool is a 6-foot (or longer) piece of ⅜₆-inch round stock bar with a small hook on one end. Its name comes from the fact that the tool can be used to enter a window and reach across the car to get to a lock or window button. But, sometimes, you use it on the same side of the car on which you inserted it. Most of those you buy come in three pieces, and you screw them together before each use. They often bend and break at the joints. If you buy one in three pieces, you should braze the pieces together. Making your own from one piece of steel is best, though.

Additional Equipment

To use those car-opening tools, you also need a flexible light and a couple of wedges. The wedges should be made of plastic, rubber, or wood. The wedges pry the door from the window to let you insert the light and the tool. The light lets you see the linkage assembly, so you can decide what tool to use and where to place it.

With most cars, many good techniques exist that let you quickly open them. A locksmith who opens a lot of cars tends to favor certain techniques and nothing is wrong with that. Whatever way gets you in quickly and professionally without damaging the vehicle is fine.

Car Parts to Reach For

Parts of the car to reach for to open it include the lock button, the bell crank, and the horizontal and vertical linkage rods.

The *bell crank* is a lever that connects to a linkage rod, which is connected to the latch or another linkage rod. One popular style of bell crank is semicircular, while another style is L-shaped.

A *horizontal rod*, as the name implies, runs parallel to the ground.

A *vertical rod* runs vertically from the lower part of the door toward the top of the door (often to a lock button).

You don't always have to use a tool within the door. Many locks are easy to impression or pick open. Standard torque wrenches used for deadbolt and key-in-knob locks don't work as well when picking a car lock. To make a better torque wrench for cars, grind the small end of a hex wrench.

Unfamiliar Car Models

When approaching an unfamiliar car model, walk around it, looking through the windows. As you walk around the car, consider the following:

1. Does it have wind wings (vent windows)?

2. Is a lock button at the top of the door?

3. Are any gaps around the doors and trunk where you may be able to insert an opening tool?

4. What type of linkage is used?

5. Can you gain access to the vehicle by removing the rear view mirror?

6. Can you manipulate the lock assembly through a hole under the outside door handle?

7. What type of pawl is used? As a rule, pre-1980 locks have free-floating pawls, and later models have rigid pawls.

Using a J Tool

If the vehicle has a lock button on top of the door, you may be able to open it with a *J* tool. First, insert a wedge between the door's weather stripping and window, to give you some space for the tool. Insert the *J* tool into the door until it passes below the window. Then, turn the tool, so its tip is under the lock button. Lift the lock button to the open position. Carefully twist the tool back into the position in which you had inserted it and remove the tool, without jerking on it, before removing the wedge.

Using a Long-Reach Tool

If you learn to use it, the long-reach tool will be one of the most useful car-opening tools you have. You can quickly unlock about 90 percent of vehicles with it, including many of the latest models. When you use this tool, it's as though you have a very long and very skinny arm. The *long-reach tool* lets you reach inside a crack of a car door to push, pull, press, and rotate knobs and buttons. You can even use it to pick up a set of keys.

To use the long-reach tool, first you place an air wedge near the top of a door to pry the door open enough to insert the tool. (Sometimes you may need to use an extra wedge.) Use a protective sleeve at the opening, and slide the tool into the sleeve. The protective sleeve is to prevent the tool from scratching the car. (You could also use cardboard or the plastic label from a bottle of soda pop.)

Most of the long-reach tools you can buy are about 56 inches long and that isn't always long enough. If you purchase a long-reach tool, get the longest one you can find. You can make your own with a 6-foot-long, ⅜-inch-diameter stainless steel rod. On one side of the rod, make a 1-inch bend at a 90-degree angle. Dip that 1-inch bend into plasti-dip or some other rubber-like coating (to give it a nonscratch coating).

Making Other Tools

You can find supplies at many hardware and home improvement stores to make your own locksmithing tools. You need flexible flat stock and bar stock of different sizes.

One tool I like a lot can be made from the plastic strapping tape used to ship large boxes. I get mine free from department stores before they throw it away. Take about 2 feet of strapping tape, fold it in half, and then glue a small piece of fine sandpaper to the center. When the glue dries, you have a nice stiff tool that can easily slide between car doors and can loop around a lock button to lift up the button. It works like a *J* tool, but it works from the top of the button. The sandpaper isn't critical, but it helps the tool grab easier.

Business Considerations

Often a person who is locked out will call several locksmiths and give the job to the one who arrives first. Or, they may get the door opened before the locksmith arrives. Either way, you may not be able to collect a fee, unless you make it clear when you received the call that you have a minimum service charge for going out on a car-opening service call.

Before working on a door, ask what attempts have been made to open the door. To improve your chance of getting an honest answer, ask in a manner that sounds as if you're just gathering technical information to help you work. If you learn that someone has been fooling with a door, don't work on that one. You don't want to be held responsible for any damage someone else may have caused.

To open a lock with vertical linkage rods, you can often use a slim jim to pull up the rod to the unlocked position, or you can use an under-the-window tool to lift the lock button. Before using an under-the-window tool on a tinted window, lubricate the tool with dishwashing liquid. This reduces the risk of scratching off the tint. You may also be able to use an *L* tool to pull up the bell crank, which is attached to the vertical linking rod.

To use a hooked horizontal rod tool, first insert a wedge between the door frame and the weather stripping, and then lower an auto light, so you can see the linkage rods. Lower the hooked end onto the rod you need, twist the tool, binding the rod, and then push or pull the rod to open the lock.

You may also want to buy a set of vent window tools for special occasions. Vent (or "wing") window opening is easy, but old weather stripping tears easily. In most cases, if the car has a vent window, it can be opened using basic vertical-linkage techniques. If you decide to use the vent window, lubricate the weather stripping with soap and water at the area where you will insert the tool. Then, take your time, and be gentle.

Special Considerations

Some models, such as the AMC Concord and Spirit, have obstructed linkage rods. It may be best to pick those locks.

Late-model cars can be tricky to work on. Many have lipped doors, which make it hard to get a tool down into the door or they have little tolerance at the gaps where you insert wedges and tools. The tight fit makes it easy to damage the car. Also, the owners may be especially watchful of any scratches you make. To reduce the risk of scratching the car, use a tool guard to cover the tool at the point it contacts the car.

Be careful when opening cars with airbags because they have wires and sensors in the door. If you haphazardly jab a tool around in the door, you may damage the system. Use a wedge and flexible light, and then make sure you can see what you're doing.

Why People Call You to Open Their Cars

A lot of people know about using a slim jim. They're sold in many hardware and auto-supply stores. And, many people know about pushing a wire hanger between the door and window to catch the lock button. People typically try those and other things before calling a locksmith. They call a locksmith because it's freezing cold, late at night, raining, or all three, and they grow tired of trying to unlock their cars themselves. Newer model cars are harder to get into using old slim jim and wire hanger techniques.

People seldom break their windows on purpose, even in emergencies. Replacing a car window is expensive and inconvenient, and there's a psychological barrier to smashing your own car window. I've been called to unlock cars with young children in them on hot days (that isn't an uncommon situation).

Car-Opening Dispatch Procedure

Having a good dispatch protocol can help you stay out of legal trouble, get the information you need to unlock the vehicle quickly, and make sure you get paid. Modify this protocol to fit your needs:

1. Speak directly to the owner or driver of the vehicle, and not to a middle-person. If the owner or driver can't come to the telephone, don't go to the job.

2. Have the person verbally confirm they want you to do the job and are authorized to hire you.

3. Always quote an estimated price (or the complete price) and a minimum service-call fee. Explain the service call fee is for the trip and it will be charged even if no other services are performed.

4. Ask how the charges will be paid (credit card, cash, or check). Explain that all charges must be paid in full and are due on your arrival.

5. Get the make, model, year, and color of the vehicle, as well as its license plate number.

6. Get the exact location of the vehicle. If the customer isn't sure, ask to speak to someone who is.

7. Get a phone number to call back, even if it's a pay phone. Tell the person someone will call back in a moment to confirm the order.

8. Call the phone number to confirm someone is really there. If no one answers, don't go to the job.

9. When you get to the job, ask to see identification, and make sure the keys are in the car. Also, have the person sign an authorization form.

10. Payment is typically paid after the automobile door is opened.

Chapter Quiz

1. Car-opening tool sets sold through locksmithing supply houses may include over 20 tools.

   A. True   **B.** False

2. You can open most cars with only five simple tools.

   A. True   **B.** False

3. The slim jim is a round piece of steel with cutouts near the bottom on both sides.

   A. True   **B.** False

4. A hooked horizontal linkage tool has a small hook on the end of the tool that lets you catch and bind a horizontal rod, and then slide it to unlock the door.

   A. True   **B.** False

5. The *J* tool goes within the door, between the window and the weather stripping, and then under the window and beneath the lock button to push the button up to the unlocked position.

   A. True   **B.** False

6. Some hooked horizontal linkage rods hook down onto the rod; others hook from the bottom of the rod.

   A. True   **B.** False

7. An *L* tool is used to push or pull on bell cranks and lock pawls.

   A. True   **B.** False

8. For versatility, you can buy or make a tool that is an *L* tool on one end and a *J* tool on the other.

   A. True   **B.** False

9. To use a hooked horizontal rod tool, first insert a wedge between the door frame and the weather stripping, and then lower an auto light, so you can see the linkage rods.

   A. True   **B.** False

10. The across-the-car tool is a 6-foot (or longer) piece of $\frac{3}{16}$-inch round stock bar with a small hook on one end.

    A. True   **B.** False

Chapter 9

SAFE BASICS

Virtually everyone has documents, keepsakes, collections, or other valuables that need protection from fire and theft. But, most people don't know how to choose a container that meets their protection needs, and they won't get much help from salespeople at department stores or home improvement centers. By knowing the strengths and weaknesses of various types of safes, you can have a competitive edge over such stores.

No one is in a better position than a knowledgeable locksmith to make money selling safes. Little initial stock is needed. They require little floor space. And safes allow for healthy price markups. This chapter provides the information you need to begin selling safes to businesses and homeowners.

Types of Safes

There are two basic types of safes: fire (or "record") and burglary (or "money"). *Fire safes* are designed primarily to safeguard their contents from fire, and *burglary safes* are designed primarily to safeguard their contents from burglary. Few low-cost models offer strong protection against both hazards. This is because the type of construction that makes a safe fire-resistant—thin metal walls with insulating material sandwiched in between—makes a safe vulnerable to forcible attacks. And, the construction that offers strong resistance to attacks—thick metal walls—causes the safe's interior to heat up quickly during a fire.

Most fire/burglary safes are basically two safes combined—usually a burglary safe installed in a fire safe. Such safes can be expensive. If a customer needs a lot of burglary and fire protection, you might suggest they buy two safes. To decide which type of safe to recommend, you need to know what your customer plans to store in the safe.

Safe Styles

Fire and burglary safes come in three basic types, based on where the safe is designed to be installed—wall, floor, and in-floor. Businesses typically use depository safes. Whatever the safe style, make sure your customer knows they should tell as few people as possible about their safe. The fewer people who know about a safe, the more security the safe provides.

Wall Safes

These types of safes are easy to install and provide convenient storage space. Unless installed in a brick or concrete block wall, such safes generally provide little burglary protection. When installed in a drywall cutout in a home, regardless of how strong the safe is, a burglar can just yank the safe out of the wall and take it with him.

Floor Safes

These safes are designed to sit on top of a floor. (Some locksmiths refer to an in-floor safe as a floor safe.) Burglary models should either be over 750 pounds or bolted into place. One way to secure a floor safe is to place it in a corner, and then bolt it to two walls and to the floor. (If you sell a large safe, make sure your customer knows the wheels should be removed.)

In-floor Safes

These safes are designed to be installed below the surface of a floor. Although they don't meet construction guidelines to earn a UL fire rating, properly installed in-floor safes offer a lot of protection

---

Installing an In-floor Safe

Although procedures differ among manufacturers, most in-floor safes can be installed in an existing concrete floor in the following way:

1. Remove the door from the safe and tape the dust cover over the safe opening.
2. At the location where you plan to install the safe, draw the shape of the body of the safe, allowing 4 inches of extra width on each side. For a square body safe, for example, the drawing should be square, regardless of the shape of the safe's door.
3. Use a jack hammer or a hammer drill to cut along your marking.
4. Remove the broken concrete and use a shovel to make the hole about 4 inches deeper than the height of the safe.
5. Line up the hole with plastic sheeting or a weatherproof sealant to resist moisture buildup in the safe.
6. Pour a 2-inch layer of concrete in the hole, and then level the concrete to give the safe a stable base to sit on.
7. Place the safe in the center of the hole and shim it to the desired height.
8. Fill the hole with concrete all around the safe and use a trowel to level the concrete with the floor. Allow 48 hours for the concrete to dry.
9. After the concrete dries, trim away the plastic and remove any excess concrete.

---

against fire and burglary. Because fire rises, a safe below a basement floor won't quickly get hot inside. For maximum burglary protection, the safe should be installed in a concrete basement floor, preferable near a corner. That placement makes it uncomfortable for a burglar to attack the safe.

Depository Safes

Depository safes, used by businesses, have a slot to allow the insertion of money. The slot prevents people from taking the money out again without using a key or combination.

Special Safe Features

Important features of some fire and burglary safes include relocking devices, hardplate, and locks. Relocking devices and hardplate are useful for a fire safe, but they are critical for a burglary safe. If a burglar attacks the safe and breaks one lock, the *relocking devices* automatically move into place to hold the safe door closed. *Hardplate* is a reinforcing material strategically located to hinder attempts to drill the safe open. Never recommend a burglary safe that doesn't have relockers and hardplate.

Safe Locks

Safe locks come in three styles: key-operated, combination dial, and electronic. The most common, *combination dial* locks, are rotated clockwise and counterclockwise to specific positions. *Electronic* locks are easy to operate and provide quick access to the safe's contents. Such locks run on batteries that must be recharged occasionally. For most residential and small business purposes, the choice of a safe lock is basically a matter of personal preference.

Underwriters Laboratories Fire Safe Ratings

Underwriters Laboratories (UL) fire safe ratings include 350-1, 350-2, and 350-4. A *350-1 UL fire safe rating* means the temperature inside the safe shouldn't exceed 350°F during the first hour of a typical home fire. A safe with a *350-2 rating* should provide such protection for up to two hours. Safes with a 350-class rating are good for storing paper documents because paper chars at 405°F. Retail prices for fire safes range from about $100 to over $4000. Most models sold in department stores and home improvement centers sell for under $300.

Underwriters Laboratories Burglary Safe Ratings

The UL 689 standard is for burglary-resistant safes. Classifications under the standard, from lowest to highest, include Deposit Safe, TL-15, TRTL-15x6, TL-30, TRTL-30, TRTL-30x6, TRTL-60, and TXTL-60. The classifications are easy to remember when you understand what the sets of letters and numbers mean. The two-set letters in a classification (TL, TR, and TX) signify the type of attack tests a safe model must pass. The first two numbers after a hyphen represent the minimum amount of time the model must be able to withstand the attack. An additional letter and number (for example, x6) tells how many sides of the safe have to be tested.

The *TL* in a classification means a safe must offer protection against entry by common mechanical and electrical tools, such as chisels, punches, wrenches, screwdrivers, pliers, hammers and sledges (up to 8-pound size), and pry bars and ripping tools (not to exceed 5 feet in length). *TR* means the safe must also protect against cutting torches.

TX means the safe is designed to protect against cutting torches and explosives.

For a safe model to earn a *TL-15* classification, for example, a sample safe must withstand an attack by a safe expert using common mechanical and electrical tools for at least 15 minutes. A *TRTL-60* safe must stand up to an attack by an expert using common mechanical and electrical tools, as well as cutting torches, for at least 60 minutes. A *TXTL-60* safe must stand up to an attack with common mechanical and electrical tools, cutting torches, and high explosives for at least 60 minutes.

In addition to passing an attack test, a safe must meet specific construction criteria before earning an Underwriters Laboratories (UL) burglary safe classification. To classify as a deposit safe, for example, the safe must have a slot or otherwise provide a means for depositing envelopes and bags containing currency, checks, coins, and the like into the body of the safe. And, it must provide protection against common mechanical and electrical tools.

The TL-15, TRTL-15, and TRTL-30 safe must either weigh at least 750 pounds or be equipped with anchors and instructions for anchoring the safe in a larger safe, in concrete blocks, or to the premises in which the safe is located. The metal in the body must be the equivalent to solid open-hearth steel at least 1-inch thick, having an ultimate tensile strength of 50,000 pounds per square inch (psi). The TRTL-15x6, TRTL-30x6, and TRTL-60 must weigh at least 750 pounds, and the clearance between the door and jamb must not exceed 0.006 inch. A TXTL-60 safe must weigh at least 1000 pounds.

TL-15 and TL-30 ratings are the most popular for business uses. Depending on the value of the contents, however, a higher rating may be more appropriate. Price is the reason few companies buy higher-rated safes. The retail price of a TL-30 can exceed $3500. A TXTL-60 can retail for over $18,000.

Such prices cause most homeowners and many small businesses to choose safes that don't have a UL burglary rating. When recommending a nonrated safe, consider the safe's construction, materials, and the thickness of door and walls. Better safes are made of steel and composite structures (such as concrete mixed with stones and steel). Safe walls should be at least ½ inch thick and the door at least 1 inch thick. Make sure the safe's bolt work and locking mechanisms provide strong resistance to drills.

Selling More Safes

You'd have to cut a lot of keys to make the money you can make from selling and installing a safe. No one is in a better position than a locksmith to sell high-quality safes. Whether you're just starting to sell them or you've been selling them for years, you can boost your sales.

The key to selling more safes is for you and all your sales staff to focus on the four Ps of marketing: products, pricing, promotion, and physical distribution.

Products

One of the most important marketing decisions you can make is which safes to stock and recommend. You need to consider quality, appearance, cost, warranty, and delivery time. Only sell good safes you believe in. Your enthusiasm for the safes you sell can make it easier for you to talk about them.

Little initial stock is needed to start selling safes. If you want to be taken seriously, however, you need to have a few on display. Most people want to see and touch a safe before buying it—much like when buying a car. Stock several sizes of each type of safe. This can make it easier for you to sell the customer up to a more expensive model.

Moving a Safe

Getting a heavy safe to your customer can be backbreaking unless you plan ahead. Consider having the safe drop-shipped, if that's an option. Most suppliers will do this for you.

As a rule of thumb, have one person help for each 500 pounds being moved. If a safe weighs more than a ton, however, use a pallet jack or a machinery mover. When moving a safe, never put your fingers under it. If the safe has a flat bottom, put three or more 3-foot lengths of solid steel rods under it to help slide the safe around, and use a pry bar for leverage.

If you're just starting to sell safes, don't stock large, heavy models because they're expensive, hard to transport, and they usually don't sell quickly. Consider stocking floor fire safes and in-floor safes. In some locales, you also may want to stock gun safes. If you're planning to sell to businesses, stock TL-rated floor safes and depository safes. Square-door safes usually sell faster than round-door models.

In addition to choosing which products to sell, you need to choose a distributor. Some distributors have a "safe dating program," in which they let you stock safes without having to pay for them until you sell them. If your distributor doesn't offer such a program, ask about stocking the safes for 90 days before paying. And see if the distributor offers training seminars.

You and your salespeople must become familiar with what you're selling. Study literature about the safes, and attend distributor and manufacturer seminars. If you don't know much about your products, potential customers will notice. As a locksmith, you're selling your expertise as well as safes. If price were the only factor, people would buy low-end safes from department stores and home improvement centers, instead of high-quality safes from you. Major manufacturers regularly offer seminars on installation and sales.

Pricing

Buy safes at good prices and sell them at a reasonable markup. Don't worry about not having the lowest prices in town. Marketing for peak profits involves adjusting the prices of products to meet the needs of customers and the needs of your company. Adjustments in prices mean adjustments in the customer's perception of prices. Prices should be based on perceived value. Many customers choose a more expensive product because they believe in the adage, "You get what you pay for."

Some customers always refuse to pay the sticker price; they feel better if they dicker the price down. You need to price your safes so all types of customers feel they're getting a good deal. One way to do this is to price most items with a little room for dickering. A good idea is to price so you can negotiate slightly—such as by giving a 2 percent discount for cash payment.

Be careful about lowering your prices, however. Every attempt should be made to sell the safe at sticker price. If the customer objects, point out the safe's benefits and features. Remember, you're selling a specialty product that protects the valuables and keepsakes of a family or business. Make it clear that you're selling a high-quality product. One major safe dealer emphasizes the importance of quality by displaying a cheap fire safe that had been broken into.

Promotion

The quality of your promotional efforts has a lot to do with how much money a customer is willing to pay for your safes. Promotion is mainly in the form of imaging and advertising. To design an effective imaging plan, you need to consider everything your customers see, hear, and smell during and after the selling process. Pay close attention to detail.

Your showroom needs to be pleasant for customers. The safes should be displayed where they can be readily seen and touched, but not where customers could trip over them. Your customers should have to walk by safes whenever they come into your shop. The display area should have good lighting, be clean (don't let dust build on the safes), and be at a comfortable room temperature. Use racks and elevated platforms, so customers don't have to bend down to touch your smaller safes.

Use plenty of manufacturer posters, window decals, and brochures in the display area. Such materials help to educate customers about your safes. Some safe manufacturers offer display materials.

In addition to placing promotional literature near the safes, include a product label on each safe. The label should include the following information: safe brand, rating, special features, warranty, regular price, sale price (if any), and delivery and installation cost. This information can help you better describe the product to customers.

A lot of locksmith shops have a web site. This can help you make direct sales, as well as promote all your products and services. You can even include a map to make it easier for customers to find you. The key to having a successful web site is to get an easily remembered domain name. If your shop name isn't already being used on the Internet, you can use it. To find out if a domain name is in use, go to www.networksolutions.com.

After getting a domain name, you can use one of the many web-site creation programs to make your web site or you can hire someone to do it. Expect to pay at least a few hundred dollars for someone to make a basic web site for you. To get ideas for creating a web site, go to an Internet search engine, such as www.hot-bot.com, and enter **lock and safe** or **safe and lock**. You'll find many locksmiths' web sites. Once you have a web site, you need to promote it by including your web address on your letterhead, business cards, service vans, and Yellow Pages ads.

The most important advertisement you can have is a listing in your local Yellow Pages. When people are looking for a safe, they don't read the newspaper, they reach for the phone book. Consider a listing under "Locks and Locksmithing" and "Safes and Vaults."

The larger your ad, the more prominent your company seems (and the more costly the ad will be). To determine the right size ad to get, look at those of your competitors. If no one else has a display ad, for instance, then don't get a display ad for that Yellow Pages heading. Instead, consider getting a bold-type listing. If many of your competitors have full- and half-page ads, however, you should have one, too—if you can afford it. If not, get the largest ad you can afford.

Some large safe dealers find television ads useful. Although advertising on national television can be expensive, advertising on local and cable television can be cost-effective. To do successful television advertising, you need to create professional-quality commercials and run them regularly. You can't simply run commercials for a month or two and expect long-term results.

The most successful safe dealers take every opportunity to talk up their safes. Whenever someone buys something at your store, ask if they need a safe. And ask during every service call. Be prepared to talk about the benefits of buying one of your safes—convenience, protection, and peace of mind. Explain that you install and service your safes. Even if the person isn't ready to buy one now, they may remember you when they are ready to buy.

Physical Distribution

The sale of a safe isn't the end of the transaction. Delivery of a safe should be done as soon as possible after the sale. Slight paranoia is a natural symptom in a customer who has just purchased a safe. If the safe takes too long to be delivered, the customer may want to cancel the order. Only work with distributors who stock a lot of safes and who can get them to you quickly.

Delivery should be done professionally and discreetly. Some safe retailers use unmarked vehicles to deliver safes. If you use an unmarked vehicle, be sure to point that out to the potential customer when you're trying to sell the safe.

By taking a little time to evaluate your current marketing strategy, you can find ways to make it better. Just remember to carefully coordinate your decisions about the four Ps of marketing and you'll improve the fifth *P*—Profits.

Chapter Quiz

1. Delivery of a safe should be done as soon as possible.

    A. True   **B.** False

2. There are two basic types of safes: fire (or "record") and burglary (or "money").

    A. True   **B.** False

3. The type of construction that makes a safe fire-resistant—thin metal walls with insulating material sandwiched in between—makes a safe vulnerable to forcible attacks.

    A. True   **B.** False

4. *Burglary safes* are designed primarily to safeguard their contents from fire.

    A. True   **B.** False

5. *Depository safes* are used by businesses and have a slot in them, so cashiers can insert money into the safes.

    A. True   **B.** False

6. Important features of fire and burglary safes include relocking devices, hardplate, and locks.

    A. True   **B.** False

7. *Hardplate* is a reinforcing material strategically located to hinder attempts to drill the safe open.

    A. True   **B.** False

8. The most common, *combination dial* models, are rotated clockwise and counterclockwise to specific positions.

    A. True   **B.** False

9. You should never recommend a burglary safe that doesn't have relockers and hardplate.

    A. True   **B.** False

10. The TL safe classification means a safe must offer protection against entry by common mechanical and electrical tools, such as chisels, punches, wrenches, screwdrivers, pliers, hammers, and pry bars, as well as ripping tools.

    A. True   **B.** False

Chapter 10

BASIC ELECTRICITY AND ELECTRONICS

This chapter covers the minimum you need to know about electricity and electronics to install electronic security and safety systems. Part 1 focuses on electricity and Part 2 deals with electronics. If you've never installed a hardwired system (or if you have, but it didn't work right), read this chapter.

Electricity

There are three types of electricity: static, direct current (DC), and alternating current (AC).

Static electricity happens in one place, instead of flowing through wires. An example is when you rub your shoes across a carpet, and then get a shock. You can also see static electricity work by rubbing a balloon on your hair, and then sticking the balloon to a wall. The most dangerous form of static electricity is lightning. For our purposes, the main concern about static electricity is preventing it from damaging electronic components.

Direct current comes from batteries and *alternating current* comes from electrical outlets in a building. Both types work by following a continuous loop from the power source through conductors (usually wires) to a load (L)—such as an alarm or other electrical device—and back again. Current from batteries flows directly from the negative terminal back to the positive. AC, on the other hand, is more erratic; it "alternates" back and forth—first in one direction, and then in another along the circuit.

AC is generated at a power plant, and then transmitted many miles through a network of high-voltage power lines. Along the route, the electricity may be more than 750,000 volts. When it gets to the substation nearest you, a transformer is used to step-down the electricity to between 5,000 and 35,000 volts. Another step-down transformer on a nearby utility pole further reduces the electricity to about 240 volts, and that's carried to your building in a cable with two separate 120-volt lines.

Typically, buildings are wired, so the two 120-volt lines work together at some outlets to provide 240 volts. The 120-volt circuits are for televisions, table lamps, and other small appliances. The 240-volt outlets are for large appliances, such as washing machines, clothes dryers, and refrigerators. (Some older homes have only one 120-volt incoming line.)

Most electronic-security components are low voltage; they require much less than 120 volts. For those, you use the transformer that comes with them. You connect the component to the transformer, and then plug the transformer into a 120-volt outlet. This reduces the 120 volts going into the transformer to an amount that's right for the electronic device. The use of low-voltage transformers makes installing electronic security systems safe.

Types of Circuits

For electricity to do useful work, it needs to flow through a circuit. A *circuit* is the pathway, or route, of electric current. A series circuit has only one pathway; it has no branches. If multiple devices are wired in series, the current flows through each in turn, and a break at any device stops the flow for all the devices. An example is the old-style Christmas tree lights. When one goes out, they all go out. A *parallel circuit* has two or more pathways for electricity to move through. If multiple devices are wired in parallel, each is wired back to the power source, so each has its own current. A *combination circuit* has both series and parallel portions.

How a Circuit Works

The three basic parts of a circuit are a power source, conductors, and a load. The cord for an appliance has at least two wires. When you plug the cord for an appliance into an electric outlet, electricity flows through the incoming wire (conductor) to the appliance (load), and then returns through the outgoing wire back to the outlet. It continues that loop until the circuit is broken.

You could turn off an appliance by using a pair of scissors and cutting one of the wires through which the electricity is flowing (but, of course, that could be dangerous). You would be breaking the circuit. When you wanted to turn the appliance on again, you would need to splice the wires back together. A safer and more convenient way of breaking a circuit is to use a switch. Whether it's a light switch, car ignition, or alarm-controller key switch, it opens and closes, or it redirects one or more circuits. From the outside, an installed switch doesn't look like much—just a little toggle, pushbutton, or turnkey. But, if you look at the back of a switch, you can see one or more tiny metal poles that move into position to complete or break circuits. The poles are little conductors that are part of the circuit. When you flip on a light switch, for example, a pole closes, to allow current to flow. When you flip the switch to the off position, the pole moves a little to prevent current flow.

Controlling Circuit Flow

Controlling the path of electricity is only part of the battle. You also need to control the current flow. If there's too little current, your components won't work correctly. Too much current and you may damage components.

A common way of describing how electricity is controlled uses the analogy of water flowing through a pipe (or hose). Water is measured in gallons and electricity in amperes (amps), symbolized by I. Water pressure is measured in pounds per square inch. Electrical pressure is measured in volts, symbolized by E. Imagine you have a 1000-foot hose attached to your kitchen faucet and you turn on the water. When you finally stretch the hose all the way out, you notice the water isn't running out very fast. To make it run faster, you could plug the hose into a fire hydrant (not recommended) for more water pressure. Or, you could use a shorter hose.

Electric current can be controlled in similar ways. You can vary the pressure (voltage), and you can change the length and thickness of the conductors. Changing conductor size is one way of changing resistance, symbolized by R, to current flow.

Going back to the water hose, you could increase or decrease resistance by adding kinks to or removing them from the hose. (I don't know why anyone would do that. I'm just trying to make a point.) *Resistance*, measured in ohms and symbolized by feet, is anything that slows the flow.

All conductors offer resistance to current flow, but some resist more than others. Copper, silver, and aluminum are good conductors, because they offer little resistance. Some materials, called insulators, are poor conductors. Examples of *insulators* include glass, dry wood, and plastic. Because plastic is a flexible insulator, it's used to sheath cable and wire to keep current from being misdirected and to prevent you from getting shocked. Electrical tape, another insulator, is used to cover breaks in the plastic insulation. Plastic connectors are used to join and insulate bare ends of wire.

Here's another way of looking at current flow. Imagine you won a million dollars, and you had to drive ten miles to pick it up. That money (or, rather, your desire for it) is the force pushing you to jump into your car, much like voltage pushes current. As you drive along, you are like current flowing. Various obstacles, such as bad weather, red lights, and police cars, are resistance. They slow you down. The fewer obstacles there are, the faster you can go (or flow). Suppose, instead of winning $1 million, you won only $10. That would be a less-motivating force, so you wouldn't drive as fast (or go as far) to get it. How far and how fast you'll go depends on your motivation and the obstacles (resistance) you face.

Ohm's Law

Having a general concept of how voltage, current, and resistance relate to one another is good, but you need to know how to put it into practical use. It would be a big hassle if every time you began an installation or were troubleshooting an electrical system, you had to resort to trial-and-error to determine what size power supply, as well as what material, length, and diameter of conductors, to use. Fortunately, you don't have to, because of a formula called Ohm's law, created by the German physicist Georg Ohm.

Ohm's law is remarkable and very simple. In a few minutes, you can have a good working knowledge of Ohm's law.

In short, Ohm's law says 1 volt can push 1 ampere through 1 ohm of resistance. Or, 2 volts can push 2 amperes through 1 ohm. This means if the resistance stays constant and you double or triple the voltage, the current also doubles or triples.

A common form of Ohm's law is E = IR, or voltage in volts equals current in amperes times resistance in ohms. (IR means *I* times *R*. The multiplication sign (×) isn't used in these kinds of equations because it can be confused with the letter *X*, which has another meaning in mathematics.)

If you know the value of any two of the variables, you can find the third variable. Variations of Ohm's law include I = E/R, or current equals voltage divided by resistance, and R = E/I, or resistance equals voltage divided by current.

Try these examples:

- If you're using a 6-volt battery in a 3-ampere circuit, how much resistance is in the circuit? To find resistance, just divide the volts by current: E/I = 6/3, or R = 2 ohms. If a circuit has 10 ohms of resistance and 5 amperes of current, how much voltage is present? To find voltage, multiply current times resistance: E = 50 volts.

- If you have a 6-volt power source connected to a 4-ohm conductor, how much current is flowing through the conductor? To find current, divide voltage by resistance: I = E/R, so E = 1.5 amperes. (If you answered 0.666 ampere, you made the mistake of dividing resistance by voltage.)

- If you have an appliance with 15 ohms of resistance plugged into a 120-volt outlet, how much current will it draw? Because I = E/R, we know that I = 8 amperes. (If you answered 0.125 ampere, you divided resistance by voltage. Try again.)

Basic Electronics

Understanding common schematic drawings can help you read electrical drawings of architects, electrical designers, and security-equipment manufacturers. You often find the drawings used in installation manuals.

Schematic Symbols

Hundreds of circuit components may go into an electronic device. To make it easier to show what components are in a device, each component is represented by a symbol, a letter, or a number. Although not every electronic technician uses the same symbols, their differences are so slight that, if you understand one, you won't likely get confused when you see another symbol for the same component.

Connecting wires (used as connectors in the circuit) are always shown as straight solid lines. They bend at sharp angles, usually 90 degrees. They rarely curve, except when crossing lines would make it unclear whether the wires are electrically connected. When no electrical connection exists between crossing lines, sometimes a half circle is drawn in the top line to show it jumping over the bottom line. That's the best way to do it. But, some people do make straight lines cross each other when the wires aren't meant to be electrically connected. When you see one straight line crossing another, you need to find out what connection is being used.

Resistors

A *resistor* is one of the most common components used in electronics: it restricts the flow of electric current and produces a voltage drop. Its value is color-coded by four bands, based on the standard of the Electronics Industries Alliance (EIA). As Table 10.1 shows, the meaning of each band depends on its position and color. The band closest to the end of the resistor represents the first position number, depending on its color. The second band represents the second position number. The third band represents the multiplier. The fourth band is used to show the tolerance of the resistor. For example, say you have a band that's marked from the position closest to the end: red, violet, orange, and silver. Using Table 10.1, you can see the first number is 2 and the second number is 7, which are the first and second position numbers, respectively. That means the number is 27. The third position number, orange, means the multiplier is 1000. When you multiply 27 by 1000, you get the nominal resistor value of 27,000 ohms. The resistor manufacturing process isn't perfect, and there's always some tolerance. The last color band tells how much above or below the nominal resistor value the actual value may be. The silver band in the last position means the actual resistor value is in the range of ±10 percent of the nominal value. That doesn't exactly mean the resistor is more or less than its nominal value—it could be exactly the same value.

There are several types of resistors. *Fixed resistors* have a fixed value in ohms. *Variable resistors* (also called rheostats or potentiometers) can be adjusted from zero to their full value to alter the amount of resistance in a circuit. *Tapped resistors* are a cross between a fixed and a variable resistor. Like a variable resistor, tapped resistors can be adjusted. Once adjusted, however, tapped resistors become like a fixed resistor and cannot be adjusted anymore.

Resistance is also shown by the letter *R*, and sequential numbers if there are multiple resistors. For example, Rl, R2, and R3 might appear at various points along a circuit diagram near schematic drawings of resistors, indicating they're the first, second, and third resistors listed on the schematic diagram. Sometimes, a resistance value is also shown on a diagram, using the Greek letter omega (Ω). For instance, you might see 30 kW near a schematic symbol for a resistor.

Table 10.1 Color-Coded Bands Based on the Standard of the Electronics Industries Alliance (EIA)

Color	Band 1	Band 2	Band 3	Band 4
Black	0	0	1	—
Brown	1	1	10	1%
Red	2	2	100	2%
Orange	3	3	1000	3%
Yellow	4	4	10,000	4%
Green	5	5	100,000	—
Blue	6	6	1,000,000	—
Violet	7	7	10,000,000	—
Gray	8	8	100,000,000	—
White	9	9	—	—
Gold	—	—	0.1	5%
Silver	—	—	0.01	10%
No color	—	—	—	20%

Capacitors

Like resistors, capacitors are among the most common components used in electronics. The *capacitor* stores electricity and acts as a filter. It allows alternating current to flow, while blocking direct current. A capacitor consists of two metal plates or electrodes separated by some kind of insulation, called dielectric, such as air, glass, mica, polypropylene film, or titanium acid barium. The type of *dielectric* used affects how much capacitance can be obtained relative to the size of the capacitor, and in which applications it can best be used.

There are two types of capacitors: fixed and variable. The tuning dial of a radio is generally attached to a fixed capacitor. When you turn the dial, you're changing its frequency. Many schematic symbols exist for identifying different types of capacitors. The letter C is used to refer to all types of capacitors, however. A whole number next to the letter C shows multiple capacitors exist and which one is within the circuit. The value, or *capacitance*, of the capacitor, shown in microfarads, may also be near the schematic symbol.

Switches

As mentioned earlier in this chapter, one way to turn a light on and off is to cut a wire that's part of the circuit, and then splice the wires back together when you want to turn the lamp on again. Using a switch is much safer and more convenient. Whenever you turn on a light or start your car, you're moving a switch into position to connect the circuit. A *switch* can have one or more contact points to allow multiple paths of current flow. Typically, switches are designed so you don't see the tiny poles that move to complete or break a circuit. If you were to take a switch apart, you would see them.

Switches are illustrated and named according to how many poles they have, and whether the poles move separately or together. A *single-pole single-throw switch* is a basic on/off type switch. Its pole makes a connection with one of two contacts. On a schematic diagram, switches are identified by the letter S. In cases of multiple switches, the S is next to a number showing its listing in the circuit.

Relays

A *relay* is a special type of switch that's operated electrically or electronically, instead of manually. A relay is like several automatic switches rolled into one. In a schematic diagram, a relay is identified by the letter *K*, and its contacts are usually numbered. The contacts are also identified as *normally open*, shown as NO, or *normally closed*, shown as NC. A normally open contact is open when no electrical power is applied. A normally closed contact is closed when no power is applied. A *solenoid* is a type of relay that uses a magnetic field to move a plunger or an arm.

Chapter Quiz

1. There are three types of electricity: static, alternating current (AC), and distant current (DC).

    A. True   **B.** False

2. *Direct current* comes from batteries.

    A. True   **B.** False

3. *Alternating current* comes from electrical outlets in a building.

    A. True   **B.** False

4. *Distant current* comes from cells and batteries.

    A. True   **B.** False

5. For electricity to do useful work, it needs to flow through a circuit.

    A. True   **B.** False

6. Most electronic security components are high voltage.

    A. True   **B.** False

7. A *circuit* is the pathway, or route, of electric current.

    A. True   **B.** False

8. A *series circuit* has only one pathway; it has no branches.

    A. True   **B.** False

9. A *parallel circuit* has two or more pathways for electricity to move through.

    A. True   **B.** False

10. A *combination circuit* has both series and parallel portions.

    A. True   **B.** False

11. The three basic parts of a circuit are a power source, conductors, and a load.

     A. True   **B.** False

12. All conductors offer resistance to current flow.

     A. True   **B.** False

13. According to Ohm's law, 1 volt can push 2 amperes through 1 ohm of resistance.

     A. True   **B.** False

14. Alternating current is generated at a power plant, and then transmitted many miles through a network of high-voltage power lines.

     A. True   **B.** False

15. A *resistor* is one of the most common components used in electronics.

     A. True   **B.** False

Chapter 11

EMERGENCY EXIT DEVICES

To comply with building and fire codes, businesses and institutions often have to keep certain doors as emergency exits, which can be easily opened by anyone at any time. (This is to help prevent not having enough quick ways out during a fire or other emergency.) In some cases, however, those doors that must remain easy to exit also need to be secured from unauthorized use (such as when the door may allow shoplifters to slip out unnoticed).

Most institutions and commercial establishments use emergency door devices as a cost-effective way to handle both matters. Such devices are easy to install and offer excellent money-making opportunities for locksmiths.

Typically, such devices are installed horizontally about 3 feet from the floor. They have a bolt that extends into the door frame to keep the door closed. They also usually incorporate either a push-bar or a clapper arm that retracts the bolt when pushed.

Some models provide outside key and pull access when an outside cylinder and door pull are installed. In these models, entry remains restricted from both sides of the door until the deadbolt is relocked by key from inside or outside the door.

Many emergency-exit door devices feature an alarm that sounds when a door is opened without a key. The better alarms are *dual piezo* (double sound). Other features to consider on an emergency-exit door device include on which hand it's installed (nonhanded models are the most versatile), the length of the bolt (a 1-inch throw is the minimum desirable length), and special security features (such as a hardened insert in the bolt).

Pilfergard Model PG-10

One of the most popular emergency exit door devices is the *Pilfergard PG-10*, which has a dual piezo alarm, can be armed and disarmed from inside or outside a door, and is easy to install on single or double doors.

The device is surface-mounted, approximately 4 to 6 feet from the floor on the interior of the door, with a magnetic actuator on the frame (or vice versa). It is armed or disarmed with any standard mortise cylinder, which is not supplied. Opening the door, removal of the cover, or any attempt to defeat the device with a second magnet, when armed, activates the alarm.

Installation

To install the Pilfergard PG-10:

1. Remove the cover by depressing the test button and lift the cover out of the slot.

2. Mark and drill holes per the template directions and drill sizes (5 for alarm unit and 2 for magnetic actuator).

3. For nonoutside cylinder installation, proceed to the next step. For outside cylinder installation, drill a 1¼-inch hole as shown on the template. Install a rim-type cylinder through the door and allow a flat-tail piece to extend 1 inch inside the door. Position the cylinder so the keyway is vertical (horizontal if the PG-10 is installed horizontally). Hold the PG-10 in position over the mounting holes and note that the outside cylinder tailpiece is centered in the clearance hole in the base of the PG-10 (if it isn't, rotate the cylinder 180 degrees). Tighten the outside cylinder-mounting screws.

4. Install the PG-10 and magnetic actuator with seven screws.

5. Install the threaded 1¼-inch mortise cylinder in the PG-10 cover, using the hardware supplied. The keyway must be horizontal, so the tailpiece extends toward the center of the unit when the key is turned.

6.   Move the slide switch to Off. Connect the battery. Hook the cover on the end slot and secure it with the two cover screws. **Note:** One of these screws acts as the tamper alarm trigger, so be sure the screws are fully seated.

Caution: When installing the PG-10 on a steel frame, it might be necessary to install a nonmagnetic shim between the magnetic actuator and the frame. This prevents the steel frame from absorbing the magnet's magnetic field, which could cause a constant alarm condition or occasional false alarms. The shim should be ½ × 2½ × ⅛–inch thick, and it may be constructed from plastic, Bakelite, or aluminum.

Testing and Operating

To test and operate the PG-10, first depress the test button with the slide switch in the Off position. Horns should sound.

To test using the magnetic actuator, proceed as follows.

1.   Close the door.

2.   Arm the PG-10 by turning the key clockwise 170 degrees.

3.   Open the door; the alarm should sound.

4.   Close the door; the alarm should continue sounding.

5.   Silence the alarm by turning the key counterclockwise until it stops.

6.   Close the door and rearm the PG-10 by turning the key clockwise until it stops.

The unit should be tested weekly using the test button to ensure the battery is working. The test button only operates when the PG-10 is turned off.

Pilfergard Model PG-2D

The *Pilfergard PG-20* is a sleek, modern version of the PG-10. The PG-20 is designed to fit all doors, including narrow stile doors. It has a flashing LED on its alarm and is as easy to install as the PG-10.

Installation

To install the PG-20, remove the cover from the mounting plate. Install the mortise cylinder (which is not supplied), keeping the key slot pointing down in the six o'clock position. Screw the lock ring on the cylinder.

Select the proper template for the specific type of door. Mark and drill ¼-inch-diameter holes per the template directions on the door and jamb. This requires four holes for the mounting plate and two holes for the magnetic actuator. **Note:** Certain narrow stile doors require only two holes for the mounting plate.

For the outside key control only: Drill a 1¼-inch-diameter hole as shown on the template.

Install a rim-type cylinder (not supplied) through the door and allow the flat tailpiece to extend ⁵⁄₁₆ inch beyond the door. Position the cylinder, keeping the key slot pointing down, in the six o'clock position. Tighten the cylinder screws.

Knock out the necessary holes from the mounting plate and install the plate on the door with the No. 8 sheet metal screws (supplied). Make sure the rim cylinder tailpiece fits in the cross slot of the ferrule if the outside cylinder is used.

Install the magnetic actuator on the door jamb with the two No. 8 sheet metal screws supplied.

Note: Installing a nonmagnetic shim between the magnetic actuator and the frame is sometimes necessary on steel frames. This prevents the absorption by the steel frame of the magnet's magnetic field, which could cause a constant or occasional false alarm. The shim should be ½ × 2½ × ⅛–inch-thick nonmagnetic material, such as plastic, Bakelite, or rubber.

Make sure the battery is connected. Cut the white jumper only on the side of the unit on which the magnet will not be installed. The unit will not function otherwise. Select the options desired as follows.

- Cut the yellow jumper for a 15-second entry delay. (The alarm will sound 15 seconds after any entry through the door if the unit is armed.) To avoid the alarm on entry, reset the unit within 15 seconds. This feature is used for authorized entrance.

- Cut the red jumper for a 15-second exit delay. (The unit will be activated after 15 seconds each time the unit is turned on with a key.) This feature allows an authorized nonalarmed exit.

- Place the shunt jumper plug in position *C* or 2. If the jumper is in position *C*, the alarm will sound continuously until the battery is discharged. If the jumper is in position 2, the alarm will be silenced after two minutes and the unit will rearm. However, the LED will continue to flash, indicating an alarm has occurred.

- Install a cover on the mounting plate, making sure the slide switch on the PC board fits into the cam hole. Secure the cover with the four screws supplied.

Testing

To field test the PG-20:

1. With the door closed, turn the key counterclockwise until it stops (less than one-quarter turn). The unit is now armed instantaneously or delayed.

2. Push the test button and the alarm should sound, verifying the unit is armed.

3. Open the door, and then close it. The alarm should sound instantaneously or delayed.

4. Note the pulsating sounder and the flashing LED, indicating an alarm.

5. To reset the unit, turn the key clockwise until it stops (less than one-quarter turn). The unit is now disarmed. This can be done at any time.

Exitgard Models 35 and 70

To install an Exitgard Model 35 or 70 panic lock:

1. With the door closed, tape the template to the inside face of the door with the center line approximately 38 inches above the floor. Mark all hole locations with a punch or awl. **Note:** The keeper is surface installed on the jamb for single doors, and on the other leaf for pairs of doors.

2.   If an outside cylinder is being used, mark the cylinder hole center. If an alarm lock pull #707 is furnished, mark the four holes on the template for the pull, in addition to the cylinder hole.

3.   Drill the holes as indicated for the lock and keeper. Drill the holes for the outside pull, if used.

4.   Loosen the hex head bolt holding the cross bar to the lock and pull the bar off the clapper arm.

5.   Remove the lock cover screws. Depress the latch bolt. Lift up and remove the lock cover.

6.   (Disregard this step if the cylinder or thumb turn is already installed.) Remove the screws holding the cylinder housing to the bolt cover. Install the rim cylinder with the keyway horizontal facing the front of the lock, opposite the slot in the rear of the cylinder support.

7.   Reinstall the support, while guiding the tailpiece into the slot cam.

8.   Use the key to try proper operation. The key should withdraw in either the fully locked or fully unlocked position of the deadbolt. If not, the cylinder and cam are mistimed. The cylinder housing must be removed, the cam turned a quarter turn to the right, and the cylinder housing reinstalled.

9.   Attach the lock to the door. For single doors, remove the keeper cover and roller, install the keeper, and then replace the roller and cover. Double-door keeper #732 is installed on the surface of the inactive leaf, as furnished.

10.   Remove the hinge pivot cover. Reinstall the push bar section on the lock clapper as far forward as it will go. Now, tighten the hex head bolt under the lock clapper. Attach the hinge-side pivot assembly, using a level or tape to assure the cross bar is level. If the cross bar is too long, loosen the hex head bolt on the underside of the clapper and remove the pivot assembly. Cut the bar to the proper length, deburr the edges, and then reinstall the pivot assembly on the cross bar. Only after the pivot base has been installed should the dogging screw be loosened, and the pivot block removed and discarded.

11.   The dogging screw should face the floor. If not, remove it and reinstall it from below. Replace the pivot cover with the hole for the dogging screw facing the floor.

12.   Test the lock operation by projecting the deadbolt by key into the keeper. Depress the cross bar fully to retract the deadbolt, and then release the latch bolt and open the door.

13.   On single doors, close the door and adjust the keeper, so the door is tightly latched. After final adjustments, install the holding screw in the keeper to maintain the position permanently. On pairs of doors, adjust the plastic slide on the keeper, so the door is tightly latched.

14.   (If you're installing Model 35, disregard this step.) Connect the power plug. Repeat Step 12. Horns should sound until the deadbolt is projected by key. Attach the self-adhesive sign to the bar and door.

15.   Replace the lock cover.

Alarm Lock Models 250, 250L, 260, and 260L

Models 250, 250L, 260, and 260L by Alarm Lock Systems, Inc. are used to provide maximum security on emergency exit doors. Each of these models features sleek architectural design and finishes, a dual piezo sounder, a low-battery alert, a simple modular construction, selectable two-minute alarm or a constant alarm, and a hardened insert in the deadbolt.

Installation

Install these models in the following way:

1. With the door closed, select the proper template. Tape it to the inside face of the door with the center line about 38 inches above the floor, according to the template directions.

2. Mark and drill the holes.

3. Remove the lock cover and four screws holding the cylinder housing to the bolt cover. Install the rim cylinder (CER) with the keyway horizontal, facing the front of the lock in the nine o'clock position. Cut the cylinder tailpiece ⅜ inch beyond the base of the cylinder. Reinstall the cylinder housing, guiding the tailpiece into the crosshole of the cam with the four screws.

4. Use the key to test for proper operation of the deadbolt. You should be able to withdraw the key from the lock in either the fully locked or the fully unlocked position of the deadbolt. If not, the cylinder and the cam are misaligned, and the cylinder housing must be removed. Turn the cam one-quarter turn to the right, and reinstall the cylinder housing. **Note:** The deadbolt can be projected into the keeper by turning the key counterclockwise. Likewise, it can be withdrawn from the keeper by turning the key clockwise one full turn.

5. For the outside cylinder only (CER-OKC): Install the rim cylinder with the keyway horizontal facing the front of the door in the three o'clock position. Use the screws supplied. Cut the tailpiece ½ inch beyond the inside face of the door.

6. For the outside cylinder only: Guide the tailpiece of the outside cylinder into the crosshole of the cam.

7. Install the lock to the door with the four screws supplied.

8. For single doors only: Remove the keeper cover, roller, and pin. Install the keeper base on the door with the two screws supplied. Reinstall the pin, roller, and cover with two screws.

9. For double doors only: Install the rub plate for a 1¾-inch-wide door from inside the door.

10. Install the keeper with two screws. Do not tighten the screws fully because the keeper will require adjustment.

11. For single doors only: Close the door, project the deadbolt, and adjust the keeper, so the door latches tightly. Retract the deadbolt, hold the keeper, release the latch, and open the door. Open the keeper cover and tighten the screws. Drill a .157-inch-diameter hole, as shown on the template, for the holding screw. Fasten the keeper with a screw. Reinstall the pin, roller, and cover with two screws.

12. For double doors only: Close the door, project the bolt, and adjust the plastic slide on the keeper, so the door is tightly latched. Tighten the screws.

Electronic Exit Lock Model 265

The Alarm Lock Systems, Inc. Model 265 emergency exit door device has the following features: a nonhanded unit, a deadbolt with a hardened-steel insert that can be operated with an outside key, a 15-second delay before the door can be opened after the clapper arm plate has been pushed, a lock that only requires 5 to 10 pounds to operate, a dual piezo horn, and a disarming beep when the bolt is retracted with a key.

Installation

To install the Model 265's lock and keeper:

1. With the door closed, select the proper template. Tape it to the inside face of the door with the center line approximately 38 inches above the floor, according to the template directions.

2. Mark and drill the following holes:

   A. For single and double doors, mark six .157-inch-diameter holes, four for the lock-mounting plate and two for the keeper.

   B. Mark a ¼-inch-diameter hole for the rub plate on double doors that are 1¾ inches thick.

   C. If an outside cylinder (CER-OKC) is used, mark the center of the 1¼-inch-diameter hole.

3. If mounting the lock on a hollow metal door and if wires are run through the door, align the lock with the holes drilled in the previous Step 2A. Mark and drill a ⅜-inch hole in the door to align with the hole in the baseplate near the terminal strip.

4. Remove the lock cover and four screws holding the cylinder housing to the bolt cover. Install rim cylinder (CER) with the keyway horizontal, facing the front of the lock in the nine o'clock position. Cut the cylinder tailpiece ⅜—inch beyond the base of the cylinder. Reinstall the cylinder housing, guiding the tailpiece into the cross-hole of the cam with four screws.

5. Use the key to test for proper operation of the deadbolt. You should be able to withdraw the key from the lock in either the fully locked or the fully unlocked position of the deadbolt. If not, the cylinder and the cam are misaligned, and the cylinder housing must be removed. Turn the cam one-quarter turn to the right, and reinstall the cylinder housing. **Note:** The deadbolt can be projected into the keeper by turning the key counterclockwise one full turn.

6. For outside cylinder only (CER-OKC): Install the rim cylinder with the keyway horizontal facing the front of the door in the three o'clock position. Use the screws supplied. Cut the tailpiece ⅜-inch beyond the inside face of the door.

7. For outside cylinder only: Guide the tailpiece of the outside cylinder into the crosshole of the outside cam.

8. Install the lock to the door with the four No. 10 screws that are supplied.

9. For single doors only: Remove the keeper cover, roller, and pin. Install the keeper base on the door with the two screws supplied. Reinstall the pin, roller, and cover with two screws.

10. Do not tighten the screws fully, because the keeper will require adjustment.

11. For single doors only: Close the door, project the deadbolt, and adjust the keeper, so the door is tightly latched. Retract the deadbolt, hold the keeper, release the latch, and open the door.

12. Open the keeper cover and tighten the screws. Drill a .157-inch-diameter hole, as shown on the template, for the holding screw, and fasten the keeper with a No. 10 screw. Reinstall the pin, roller, and the cover with two screws.

13. For double doors only: Close the door, project the bolt, and adjust the plastic slide on the 732 keeper, so the door is tightly latched. Tighten the screws.

14. A fine adjustment in the latch and electromagnet mechanism might be necessary. With the door pulled fully closed, check to make certain the backstop is in complete contact with the electromagnet.

15.   A small gap, approximately ½ inch or less, will be between the rod and latch. If not, loosen the Allen head screw and slide the electromagnet to the right or left until it's adjusted.

16.   Retighten the Allen head screw.

Installing the Control Box

To install the control box for Model 265:

1.   Remove the control box cover.

2.   Select a location for the control box on the hinge side of the door and mount it to the wall, using the three No. 10 × ¾-inch self-tapping screws.

Wiring

Model 265 is wired in the following way:

1.   A four-conductor No. 22-AWG cable is needed to connect the control box to the lock. You can bring electric current from the hinge side of the door frame to the door in two ways. First, use the Armored Door Loop Model 271 and disconnect one of the 271 end boxes. Insert the loose end of the armored cable into the ½-inch hole on the control box and secure it with the retaining clip. Or, second, you can use a continuous conductor hinge with flying leads.

2.   Connect one end of the four-conductor cable to the control box terminal strip P2 and the other end of the cable to the lock terminal strip P3. The terminal strips are marked as follows.

   • (1) SEN –sence (2) EM electromagnet (3) +9VDC (4) –ground. **Note:** Do not cross wires.

3.   Connect one end of an approved twin lead cable to the terminal strip at Pl-1 and Pl-2, and connect the other end of the cable to the transformer provided. Do not plug in the transformer at this time.

4.   Connect one end of another approved twin lead cable to the terminal strip at Pl-3 and Pl-4. Connect the other end to the normally closed alarm relay contacts of either an approved supervised automatic fire detection system or an approved supervised automatic sprinkler system.

5.   Install the control box cover.

Operating

Before installing the lock cover:

1.   For a continuous alarm, leave the black jumper plug on the terminal strip installed. For a two-minute alarm shutdown, remove the black jumper plug from the terminal strip.

2.   Connect the 9-volt battery to its connector. A short beep will sound, ensuring the 265 is powered and ready.

3.   Plug the transformer into a continuous 115 AC volt source. **Note:** The red pilot lamp on the control box is lit.

4.   Open the door and install the lock cover with the four screws supplied.

Testing

To test the 265, lock and unlock the door using the cylinder key. **Note:** A small beep sounds when the bolt is retracted, indicating a disarmed condition. The door can be opened by pushing the clapper plate, after a 15-second wait, without causing an alarm.

Lock the door again and push the clapper plate. Immediately the alarm will pulse loudly and the 265 latch will impede the opening of the door for approximately 15 seconds, and then remain unimpeded until it is manually reset with the key. If continuous alarm was selected, the piezo sounder will remain on for two minutes, and then reset.

In the event of a power failure, the impeding latch will be disabled, but the 9-volt battery will provide standby power for the alarm circuit. This can be tested by disconnecting the transformer from the 115-volt source and pushing the clapper plate. The door should now open immediately and the alarm should sound.

In the event of a fire panel alarm, the impeding latch will also be disabled. This can be tested by disconnecting the wire going to the control box at Pl-3 and pushing the clapper plate. The door should open immediately and the alarm should sound.

Test the unit's battery by pushing and holding the test button on the control box, and then pushing the clapper plate to alarm the unit. If the piezo sounder is weak or doesn't operate at all, replace the 9-volt battery.

Alarm Lock Models 700, 700L, 710, and 710L

Standard features for each emergency exit door device in Alarm Lock Systems, Inc. 700 and 710 Series include the following: a nonhanded unit, a deadbolt with a hardened steel insert that can be operated by an outside key and can be used for single or double doors, a deadlatch for easy access from inside without an alarm, a loud dual piezo horn, a selectable continuous or two-minute alarm shutdown, a disarming beep when the bolt is retracted with a key, a low-battery beep when the battery needs to be replaced, and a retriggerable alarm after a two-minute shutdown (for Model 710 only).

Installation

The 700 and 710 Series models are installed as follows:

1. With the door closed, select the proper template. Tape it to the inside face of the door with the center line approximately 38 inches above the floor, according to template directions.

2. Mark and drill the following holes (see the template for details):

    - For single and double doors, mark six .157-inch-diameter holes: four for the lock mounting plate and two for the keeper.

    - Mark a ½-inch-diameter hole for the rub plate on double doors 1¾ inches thick.

    - If the outside cylinder (CER-OKC) is used, mark the center of the ¼-inch-diameter holes.

    - If the outside pull Model 707 is used, mark the center of the four ¼-inch-diameter holes.

    - If you're mounting the lock on a hollow metal door and if wires are to be run through the door, drill hole X (see the template). **Note:** If the outside pull is used, drill four ¼-inch-diameter holes through the door from the inside. Then, drill ¾-inch-diameter holes 1¼ inches deep from the outside of the door.

3. Remove the lock cover and four screws holding the cylinder housing to the bolt cover.

4. Install the rim cylinder (CER) with the keyway horizontal, facing the front of the lock in the nine o'clock position. Cut the cylinder tailpiece ⅜ inch beyond the base of the cylinder. Reinstall the cylinder housing, guiding the tailpiece into the crosshole of the cam with the four screws.

5. Use the key to test for proper operation of the deadbolt. You should be able to withdraw the key from the lock in either the fully locked or fully unlocked position of the deadbolt. If not, the cylinder and the cam are misaligned, and the cylinder housing must be removed. Give the cam one-quarter turn to the right, and reinstall the cylinder housing. **Note:** The deadbolt can be projected into the keeper by turning the key counterclockwise, and it can be withdrawn from the keeper by turning the key clockwise one full turn.

6. For the outside cylinder only (CER-OKC): Install the rim cylinder with the keyway horizontal facing the front of the door in the three o'clock position, using the screws supplied. Cut the tailpiece ½ inch beyond the inside face of the door.

7. For the outside cylinder only: Guide the tailpiece of the outside cylinder into the crosshole of the outside cam.

8. Install the lock loosely to the door with four No. 10 screws as supplied. Do not tighten them at this time.

9. Insert the bar and the channel assembly under the channel retainer bracket, which is mounted to the lock baseplate. Hold the bar and channel assembly horizontally against the door using a level.

10. Slide the end-cap bracket into the end of the channel and, using the bracket as a template, mark and drill the two .157-inch-diameter mounting holes on the door. If the channel is too long, cut the channel and the channel insert to the proper length and deburr the edges.

11. Attach the push bar to the lock at the clapper-arm hinge bracket, using the ½-inch screw and No. 10 internal tooth lockwasher provided.

12. Mount the end-cap bracket to the door with the two No. 10 screws provided, and then tighten the lock securely to the door.

13. Attach the end cap to the end-cap bracket using the ½-inch oval head screw provided.

14. For single doors only: Remove the keeper cover, roller, and pin. Install the keeper base on the door with the two screws supplied. Reinstall the pin, roller, and cover with the two screws.

15. For double doors only: Install the rub plate for a 1¾-inch-wide door from inside the door. Also install the 732 keeper with the two No. 10 screws supplied. Do not tighten the screws fully because the keeper will require adjustment.

16. For single doors only: Close the door, project the deadbolt, and adjust the keeper, so the door is tightly latched. Retract the deadbolt, hold the keeper, release the latch, and then open the door.

17. Open the keeper cover and tighten the screws. Drill a .157-inch-diameter hole, as shown on the template, to hold the screw. Fasten the keeper with a No. 10 screw. Reinstall the pin, roller, and cover with two screws.

18. For double doors only: Close the door, project the bolt, and adjust the plastic slide on the 732 keeper, so the door is tightly latched, and then tighten the screws.

Operating 700 and 710 Series Models

Before installing the lock cover:

1. For a continuous alarm, leave the black jumper plug on the terminal strip, as installed.

2. For a two-minute auto-alarm shutdown, remove the black jumper plug from the terminal strip.

3. Connect the battery connector to the 9-volt battery, observing the proper polarity. A short beep will sound, ensuring the lock is powered and ready.

4. Install the lock cover with the four screws supplied.

5. Close the door.

Testing

To test a 700 or 710 Series model:

1. Lock and unlock the door using the cylinder key. **Note:** A small beep sounds when the deadbolt is retracted, indicating a disarmed condition. The door can now be opened without an alarm, by pushing the push bar.

2. Lock the door again and push the push bar to open it. Immediately, the alarm will pulse loudly and, if the continuous alarm was previously selected, the alarm will sound until the lock is manually reset by locking the deadbolt with the key. If the auto-alarm shutdown was selected, the alarm will sound for two minutes, and then reset.

3. When the battery becomes weak, the sounder will emit a short beep approximately once a minute, indicating the battery needs replacing.

Special Operations

If the unit has a retriggerable alarm (Model 260), after the initial two-minute alarm and auto shutdown, the alarm will retrigger if the door is opened again. This function remains retriggerable until the door is reclocked with the key. Whenever an alarm is caused by opening the door and the door is left open, the two-minute alarm shutdown will be inhibited.

To use the unit's dogging operation, insert the ³⁄₁₆-inch Allen wrench into the dogging latch through the hole in the channel insert. Turn the dogging latch counterclockwise a half turn, push in the push bar, and turn the dogging latch clockwise a quarter turn until it stops. Release the push bar and notice that it stays depressed and the door is unlatched.

Alarm Lock Model 715

Standard features of the Model 715 include: the unit is nonhanded; a deadbolt with a hardened steel insert that can be operated by an outside key; a 15-second delay before the door can be opened after the push bar has been pushed; a lock that only requires 5 to 10 pounds of force to operate and that can be used for single or double doors; loud dual piezo horn; selectable, continuous, or two-minute alarm shutdown; and a disarming beep when the bolt is retracted with a key.

Installation

The Model 715 electronic exit lock is installed in the following way:

1. With the door closed, select the proper template and tape it to the inside face of the door, with the center line approximately 38 inches above the floor.

2. Mark and drill the following holes (see the template for details):

    A. For single and double doors: Mark six .157-inch-diameter holes, four for the lock mounting plate and two for the keeper.

    B. Mark a ¼-inch-diameter hole for the rub plate on double doors 1¾ inches thick.

    C. If an outside cylinder (CER-OKC) is used, mark the center of the 1¼-inch-diameter hole.

    D. If you're mounting the lock on a hollow metal door and if wires are to run through the door, align the lock with the holes drilled in Step 2A. Mark and drill a ⅜-inch hole in the door to align it with the hole in the baseplate near the terminal strip P3.

3. Remove the lock cover and the four screws holding the cylinder housing to the bolt cover.

4. Install the rim cylinder (CER) with the keyway horizontal, facing the front of the lock in a nine o'clock position. Cut the cylinder tailpiece ⅜ inch beyond the base of the cylinder. Reinstall the cylinder housing, guiding the tailpiece into the crosshole of the cam with four screws.

5. Use the key to test for proper operation of the deadbolt. You should be able to withdraw the key from the lock in either the fully locked or the fully unlocked position of the deadbolt. If not, the cylinder and the cam are misaligned, and the cylinder housing must be removed. Turn the cam a quarter turn to the right, and then reinstall the cylinder housing.

    Note: The deadbolt can be projected into the keeper by turning the key counterclockwise, and it can be withdrawn from the keeper by turning the key clockwise one full turn.

6. For the outside cylinder only (CER-OKC): Install the rim cylinder with the keyway horizontal facing the front of the door in the three o'clock position. Use the screws supplied. Cut the tailpiece ⅜ inch beyond the inside face of the door.

7. For the outside cylinder only: Guide the tailpiece of the outside cylinder into the crosshole of the outside cam.

8. Install the lock loosely to the door with the four No. 10 screws supplied. Do not tighten the screws at this time.

9. Insert the bar and channel assembly under the channel retainer bracket, which is mounted to the lock baseplate. Using a level, hold the bar and channel assembly horizontally against the door.

10. Slide the end-cap bracket into the end of the channel and, using the bracket as a template, mark and drill the two .157-inch-diameter mounting holes on the door. If the channel is too long, cut the channel and the channel insert to the proper length and deburr the edges.

11. Attach the push bar to the lock at the clapper-arm hinge bracket using the ½-inch screw and No. 10 internal tooth lockwasher provided.

12. Mount the end-cap bracket to the door with the two No. 10 screws provided, and then tighten the lock securely to the door.

13. Attach the end cap to the end-cap bracket using the ½-inch oval head screw provided.

14. For single doors only: Remove the keeper cover, roller, and pin. Install the keeper base on the door with the two screws supplied. Reinstall the pin, roller, and cover with the two screws.

15. For double doors only: Install the rub plate for a 1¾-inch-wide door, from inside the door. Also install the 732 keeper with the two No. 10 screws. Do not tighten the screws fully because the keeper will require adjustment, as mentioned in the next step.

16. For single doors only: Close the door, project the deadbolt, and adjust the keeper, so the door is tightly latched. Then, retract the deadbolt, hold the keeper, release the latch, and open the door.

17. Open the keeper cover and tighten the screws. Drill a .157-inch-diameter hole, as shown on the template, for the holding screw, and fasten the keeper with a No. 10 screw. Reinstall the pin, roller, and cover with the two screws.

18. For double doors only: Close the door, project the bolt, and adjust the plastic slide on the 732 keeper, so the door is tightly latched. Tighten the screws.

Installing the Control Box

To install the control box for Model 715, remove the control box cover. Select a location for the control box on the hinge side of the door and mount it to the wall, using the three ¾-inch self-tapping screws.

Wiring

Model 715 is wired in the following way:

1. A four-conductor No. 22-AWG cable is needed to connect the control box to the lock. You can bring electric current from the hinge side of the door frame to the door in two ways. One, use the Armored Door Loop Model 271 by disconnecting one of the 271 end boxes and inserting the loose end of the armored cable into the ½-inch hole on the control box. Secure it with the retaining clip. Or, second, use a continuous conductor hinge with flying leads.

2. Connect one end of the 4-conductor cable to the control box terminal strip P2 and the other end of the cable to the lock terminal strip P3. The terminal strips are marked as follows:

   • (1) SEN –sence (2) EM –electromagnet (3) +9 VDC (4) –ground. **Note:** Do not cross wires.

3. Connect one end of an approved twin lead 18-2 cable to the terminal strip at Pl-1 and Pl-2, and then connect the other end of the cable to the 12 Vac 20VAC transformer provided. Do not plug in the transformer at this time.

4. Connect one end of another approved twin lead 22-2 cable to the terminal strip at Pl-3 and Pl-4; connect the other end to the normally closed alarm relay contacts of either an approved supervised automatic fire-detection system or an approved supervised automatic sprinkler system.

5. Install the control box cover.

Operating

Before installing the lock cover, do the following:

1. For a continuous alarm, leave the black jumper plug on the terminal strip, as installed.

2. For a two-minute auto-alarm shutdown, remove the black jumper plug from the terminal strip.

3. Connect the 9-volt battery to its connector. A short beep will sound, ensuring the 715 is powered and ready.

4. Plug the transformer into a continuous 115 AC volt source. **Note:** The red pilot lamp on the control box is lit. Open the door and install the lock cover with the four screws supplied.

Testing

To test a Model 715:

1. Lock and unlock the door using the cylinder key. Notice the small beep when the deadbolt is retracted, indicating a disarmed condition. The door can be opened by pushing the push bar, after a 15-second wait, without causing an alarm.

2. Lock the door again and push the push bar. Immediately the alarm will pulse loudly and the 715 latch will impede the opening of the door for approximately 15 seconds, and then remain unimpeded until it is manually reset with the key. Or, if the continuous alarm was selected above, the piezo sounder will remain on for two minutes, and then reset.

3. In the event of a power failure, the impeding latch will be disabled, but the 9-volt battery will provide standby power for the alarm circuit. Test this by disconnecting the transformer from the 115-volt source and pushing the push bar. The door should now open immediately and the alarm should sound.

4. In the event of a fire panel alarm, the impeding latch will also be disabled. Test this by disconnecting the wire going to the control box Pl-3 and pushing the push bar. The door should open immediately and the alarm should sound.

Test the battery by pushing and holding the test button on the control box and pushing the push bar to alarm the unit. If the piezo sounder sounds weak or doesn't operate at all, replace the 9-volt battery.

Chapter Quiz

1. To comply with building and fire codes, businesses and institutions often have to keep certain doors as emergency exits, which can be easily opened by anyone at any time.

   A. True   **B.** False

2. Typically, emergency exit-door devices are installed horizontally about 3 feet from the floor.

   A. True   **B.** False

3. Some exit-door devices provide outside key and pull access when an outside cylinder and door pull are installed.

   A. True   **B.** False

4. Many emergency exit-door devices feature an alarm that sounds when a door is opened without a key.

   A. True   **B.** False

5. Exit devices now come in a wide range of styles (such as push bar and cross bar) and finishes (such as aluminum, stainless, brass, and bronze).

   A. True   **B.** False

Chapter 12

WIRELESS AND HARDWIRED ALARMS

N owhere have recent advances in electronic and computer technology been more apparent than with security and home automation systems. Many types of systems that sell for under $1,000 today weren't available ten years ago at any price, and some of today's lowest priced systems are more effective and more reliable than ever.

To get your money's worth, however, you have to know what to look for. This chapter reviews a wide range of electronic security systems and devices. I explain why some of them can be useful and why many others can be costly nuisances. I also show you the basic installation procedure used for many types of alarms and home automation systems.

Intruder Alarms

More than 600 inmates of an Ohio prison were asked what single thing they would use to protect their homes from burglars. The most popular choice was a dog; the next was a burglar alarm. Other studies show that many police officers also believe a burglar alarm can make a home safer.

I favor installing intruder alarms, but they're not useful for everyone. To benefit from a burglar alarm, you and everyone in your home must learn how to operate it properly and must use it consistently. Everyone must remember to keep all windows and doors of the house closed when the system is armed. Many homeowners pay thousands of dollars for an alarm system only to discover that using it is too much trouble.

Tricks of Hardwiring

Although hardwired systems generally are more reliable and less expensive than their wireless counterparts, few laypersons like to install hardwired alarms. Sometimes, getting a length of wire from a control panel to the sensors can be tricky. Here are some tips that might help:

- When running wire from one floor to another, try using the existing openings used by plumbing or vents.
- If you have to drill a hole to get wire from one floor to another, consider drilling in a closet or another place that won't be noticeable. As a last resort, consider drilling as close to a corner as possible.
- Try running wire above drop ceilings.
- Try running wire under wall-to-wall carpet as close to the walls as possible (not in high-traffic pathways).
- If you can't hide the wire you're running, consider running it through plastic strips of conduit. (Conduit not only makes the run look neater, but it also protects the wire.)
  - If you can't hide the wire and you aren't using conduit, try to run the wire close to the baseboard.
- When running wire without conduit, you may need to staple the wire. Use rounded staples only. Flat-back staples may cut into the wire and cause problems.

Contrary to popular belief, a burglar alarm doesn't stop or deter burglars. It only warns of their presence (if it's turned on during a break-in). Some burglar-alarm sellers say that if you have an alarm, it will make burglars think twice about trying to break into your home. Actually, it isn't having the alarm that deters intruders; it's their belief that a home or office has an alarm that will stop most of them. Often the only part of a burglar alarm that can be seen from outside is the window sticker. If you use an alarm system window stickers and yard sign, few burglars will know whether you do or don't have an intruder alarm.

Alarm systems are sold as complete kits or you can sell the components separately. The components are likely to include a control panel, a siren or bell, and various detection devices.

Detection devices (or sensors) are the eyes and ears of the system. They sense the presence of an intruder and relay the information to the control panel, which activates the siren or bell. Today, you have more detection devices to choose from than ever before, but if you choose the wrong ones or install them in the wrong place, you'll have a lot of false alarms or a system that doesn't detect an intruder.

Some detection devices respond to movement, some to sound, and others to body heat. The principle behind each is similar. When an alarm system is turned on, the devices sense a "normal" condition. When someone enters a protected area, the devices sense a disturbance in the normal condition and trigger an alarm.

Most detection devices fall within two broad categories: perimeter and interior. *Perimeter devices* are designed to protect a door, window, or wall. They detect an intruder before entry into a room or building. The three most common perimeter devices are foil, magnetic switches, and audio discriminators. *Interior* (or space) *devices* detect an intruder on entry into a room or protected area. The five most common interior devices are ultrasonic, microwave, passive infrared, quad, and dual-tech detectors.

Foil

You've probably seen foil on storefront windows. It's a thin, metallic, lead-based tape, usually ½- to 1-inch wide, that's applied in continuous runs to glass windows and doors. Sometimes foil is used on walls. Like wire, *foil* acts as an electrical conductor to make a complete circuit in an alarm system. When the window (or wall or door) breaks, the fragile foil breaks, creating an incomplete circuit and triggering the alarm.

Usually foil comes in long, adhesive-backed strips and is applied along the perimeter of a sheet of glass or dry wall. Each end of a run must be connected to the alarm system with connector blocks and wire. Foil is popular in stores because it costs only a few cents per foot and acts as a visual deterrent.

Foil has three major drawbacks:

1. It can be tricky to install properly.

2. It breaks easily when a window is being washed.

3. Many people consider it unsightly.

Whether or not you like foil, foil alone is rarely enough to protect a home. Other detection devices also should be used.

Magnetic Switches

The most popular type of perimeter device is the *magnetic switch*, which is used to protect doors and windows that open. Magnetic switches are reliable, inexpensive, and easy to install.

As its name implies, the magnetic switch consists of two small parts: a magnet and a switch. Each part is housed in a matching plastic case. The switch contains two electrical contacts and a metal spring-loaded bar that moves across the contacts when magnetic force is applied. When magnetic force is removed, the bar lifts off one of the contacts, creating an open circuit and triggering an alarm condition.

In a typical installation, the magnet is mounted on a door or window, and the switch is aligned about ½-inch away on the frame. When an intruder pushes the door or window open, the magnet is moved out of alignment. Some magnetic switches are rectangular, for surface mounting. Others are cylindrical, for recessed mounting in a small hole. The recess-mounted types look nicer because they're less conspicuous, but they're a little harder to install.

One problem with some magnetic switches is that an intruder can defeat them by using a strong magnet outside a door or window to keep the contacts closed. Some models can be defeated by placing a wire across the terminal screws of the switch, jumping the contacts. Another problem is this: if a door is loose fitting, the switch and magnet can move far enough apart to cause false alarms.

Wide-gap reed switches can be used to solve those problems. Because reed switches use a small reed instead of a metal bar, they're less vulnerable to being manipulated by external magnets. The wide-gap feature allows a switch to work properly even if the switch and magnet move from 1 to 4 inches apart. Some magnetic switches come with protective plastic covers over their terminal screws. The covers thwart attempts at jumping. Most types of magnetic switches cost just a few dollars each.

Audio Discriminators

Audio discriminators trigger alarms when they sense the sound of glass breaking. The devices are very effective and easy to install. According to a survey by *Security Dealer* magazine, over 50 percent of professional alarm installers favor audio discriminators over all other forms of glass break-in protection.

By strategically placing audio discriminators in a protected area, you can protect several large windows at once. Some models can be mounted on a wall up to 50 feet away from the protected windows. Other models, equipped with an omnidirectional pickup pattern, can monitor sounds from all directions and are designed to be mounted on a ceiling for maximum coverage.

A problem with many audio discriminators is they confuse certain high-pitched sounds—such as keys jingling—with the sound of breaking glass and produce false alarms. Better models require both the sound of breaking glass and shock vibrations simultaneously to trigger their alarm. This feature greatly reduces false alarms.

Another problem with audio discriminators is their alarm is triggered only if glass is broken. An intruder can bypass the device by cutting a hole through the glass or by forcing the window sash open. Audio discriminators work best when used in combination with magnetic switches.

Ultrasonic Detectors

Ultrasonic detectors transmit high-frequency sound waves to sense movement within a protected area. The sound waves, usually at a frequency of over 30,000 hertz, are inaudible to humans, but can be annoying to dogs. Some models consist of a transmitter that is separate from the receiver, while others combine the two in one housing.

In either type, the sound waves are bounced off the walls, floor, and furniture in a room until the frequency is stabilized. Thereafter, the movement of an intruder causes a change in the waves and triggers the alarm.

A drawback to ultrasonic detectors is they don't work well in rooms with wall-to-wall carpeting and heavy draperies because soft materials absorb sound. Another drawback is ultrasonic detectors do a poor job of sensing fast or slow movements and movements behind objects. An intruder can defeat a detector by moving slowly and hiding behind furniture. Ultrasonic detectors are prone to false alarms caused by noises, such as a ringing telephone or jingling keys. Although they were very popular a few years ago, ultrasonic detectors are a poor choice for most homes. They can cost over $60; other types of interior detectors cost less and are more effective.

Microwave Detectors

Microwave detectors work like ultrasonic detectors, but they send high-frequency radio waves instead of sound waves. Unlike ultrasonic waves, these microwaves can go through walls and be shaped to protect areas of various configurations. Microwave detectors are easy to conceal because they can be placed behind solid objects. They are not susceptible to loud noises or air movement when adjusted properly.

The big drawback to microwave detectors is their sensitivity makes them hard to adjust properly. Because the waves penetrate walls, a passing car can prompt a false alarm. Their alarms also can be triggered by fluorescent lights or radio transmissions. Microwave detectors are rarely useful for homes.

Passive Infrared Detectors

Passive infrared (PIR) detectors became popular in the 1980s. Today, they are the most cost-effective type of interior device for homes. A *PIR detector* senses rapid changes in temperature within a protected area by monitoring infrared radiation (energy in the form of heat). A PIR detector uses less power, is smaller, and is more reliable than either an ultrasonic or a microwave detector.

The PIR detector is effective because all living things give off infrared energy. If an intruder enters a protected area, the device senses a rapid change in heat. When installed and adjusted properly, the detector ignores all gradual fluctuations of temperature caused by sunlight, heating systems, and air conditioners.

A typical PIR detector can monitor an area measuring about 20 by 30 feet or a narrow hallway about 50 feet long. It doesn't penetrate walls or other objects, so a PIR detector is easier to adjust than a microwave detector. Also, it doesn't respond to radio waves, sharp sounds, or sudden vibrations.

The biggest drawback to PIR detectors is they don't "see" an entire room. They have detection patterns made up of "fingers of protection." The spaces outside and between the fingers aren't protected by the PIR detector. How much of an area is monitored depends on the number, length, and direction of zones created by a PIR detector's lens and on how the device is positioned.

Many models have interchangeable lenses that offer a wide range of detection pattern choices. Some patterns, called *pet alleys*, are several feet above the floor to allow pets to move about freely without triggering the alarm. Which detection pattern is best for you depends on where and how your PIR detector is being used.

A useful feature of the latest PIR detectors is *signal processing* (also called *event verification*). This high-tech circuitry can reduce false alarms by distinguishing between large and small differences in infrared energy.

Quads

A *quad PIR detector* (or *quad*, for short) consists of two dual-element sensors in one housing. Each sensor has its own processing circuitry, so the device is basically two PIR detectors in one. A quad reduces false alarms because, to trigger an alarm, both PIR detectors must detect an intrusion simultaneously. This feature prevents the alarm from activating in response to insects or mice. A mouse, for example, may be detected by the fingers of protection of one of the PIR detectors, but it would be too small to be detected by both at the same time.

Dual Techs

Detection devices that incorporate two different types of sensor technology into one housing are called *dual-technology devices* (or *dual techs*). A dual tech triggers an alarm only when both technologies sense an intrusion. Dual techs are available for commercial and residential use, but because they can cost several hundred dollars, dual techs are used more often by businesses. The most effective dual tech for homes is one that combines PIR detectors and microwave technology.

For this type of dual tech to trigger an alarm, a condition must exist that simultaneously triggers both technologies. The presence of infrared energy alone, or of movement alone, would not trigger an alarm. Movement outside a wall, which ordinarily might trigger a microwave, for example, won't trigger a dual tech because the PIR element wouldn't simultaneously sense infrared energy.

Chapter Quiz

1. A quad PIR detector consists of four dual-element sensors in one housing.

   A. True   **B.** False

2. Detection devices that incorporate four different types of cylinders in one housing are called dual-technology devices.

   A. True   **B.** False

3. A big drawback to PIRs is they monitor an entire room and go through walls.

   A. True   **B.** False

4. The PIR detector is effective because no living things give off infrared energy.

   A. True   **B.** False

5. Microwave detectors work like PIRs.

   A. True   **B.** False

Chapter 13

HOME AUTOMATION

Although locks, light, sound, and other elements play a part in home and office security and safety, each of these elements must be controlled separately in most places. By using a home automation system, however, you can make several or all of these systems in a home work automatically to provide more security, safety, and convenience.

"Home automation" is a generic term that refers to any automated technology used in homes—such as automated lights that come on when someone pulls into your driveway. If the right attachments are used, all home automation systems can perform many of the same functions. However, important differences exist among the three basic types of systems.

Programmable Controller

A more versatile type of home automation system is one that uses a programmable controller, which is integrated into your home's electrical power line. A programmable controller allows all your automation devices to work together under a central control. By touching a keypad in your bedroom, for instance, you could turn down the heat in your home, arm your burglar alarm, and turn on your outdoor lights. Or, you could use your programmable controller to make all those things occur automatically every night at a certain time. However, it can cost up to $20,000 to have a full-blown, power-line system installed in a home.

Smart House Integrated System

One of the latest and most sophisticated home automation systems is the Smart House. Although the term "smart house" sometimes is used to refer to a wide variety or a combination of home automation systems, it's actually a brand name for a unique system of automating a home. The *Smart House* integrates a unique wiring system and computer-chip language to allow all the televisions, telephones, heating systems, security systems, and appliances in a home to communicate with each other and to work together.

If your customer's refrigerator door has been left open, for instance, the Smart House could signal their television set to show a picture of a refrigerator in the corner of the screen until the door is closed. A smoke detector in the Smart House could signal your heating system to shut down during a fire. Its communication ability is one of the most important differences between the Smart House and all other home automation systems. The basic installation cost of a Smart House system is about the same as that of a power-line system, but with a Smart House system, you also may need to purchase special appliances.

How the Smart House System Works. To understand how Smart House technology works, it's important to realize that the technology was the result of a joint effort among many appliance manufacturers, security system manufacturers, and home building and electronics trade associations. All of them agreed on standards that allow special appliances and devices to work in any Smart House. A Smart House uses a system controller, instead of a fuse panel, and Smart blocks, instead of standard electrical outlets. Appliances designed to work in a Smart House are called *Smart appliances* and all of them can be plugged into any Smart block. The same Smart block into which you plug your television, for instance, can be used for your telephone or coffee pot. When a Smart appliance is plugged into a Smart block, the system controller receives a code to release power, and it coordinates communication between that appliance and the other Smart appliances.

A big difference between standard electrical outlets and Smart blocks is this: electricity is always present in the standard outlets. If you were to stick a metal pin into one of your standard outlets, you would get an electric shock. If you were to stick a pin into a Smart block, you wouldn't get shocked because no electricity would be present. Only a device that has a special computer-chip code can signal the Smart House system controller to release electricity to a particular Smart block—unless you override the signal.

With a Smart House, you have the option of programming any or all of the Smart blocks to override their need for a code. Such an option enables you to use standard appliances in your Smart blocks, in much the same way you use your electrical outlets now. Standard appliances can't communicate with each other or with Smart appliances. You might want to override a Smart block if some of your appliances aren't Smart appliances.

Because the Smart House is a new technology, few Smart appliances are available. If the technology becomes more widely used, the demand for Smart appliances will increase.

Home Automation Controllers

With either a Smart House or a power-line system, you need only one controller to make the system do anything you want it to do. For convenience, however, you might want controllers installed at several locations in your home. In addition to a keypad, you can use your telephone, a computer, or a touch screen for remote control of your system. A touch screen looks like a large television that is mounted into a wall. It displays a "menu" of your options—lighting, security, audio, video, temperature controls, and so on—and you can make your selection just by touching the screen. If you were to touch "Security," for example, a blow-up of the floor plan of your home may appear on the screen. You would be able to see whether any windows or doors are open, whether your alarm system is on or off, and other conditions related to your home's security. You also would be able to secure various areas of your home just by touching the screen.

X-10 Compatible Home Automation Systems

You may think a comprehensive electronic security system costs thousands of dollars and requires a professional to install it. Some security systems do, but companies such as IBM, Leviton, RCA, Heath, Radio Shack, Sears, and Stanley offer effective, low-cost, home-security products that are X-10-compatible. And X-10-compatible devices are easy for security professionals to install.

They all share the same X-10 technology—a system that enables security and home automation components to operate using house wiring and compatible radio frequencies. This means you can mix components from several manufacturers. Best of all, you can create an effective X-10 automation and security system for less than $200 and expand it later by selecting from a wide variety of components.

With an automation and security system in place, you can operate house lights and appliances or trigger the alarm siren with a hand-held remote control device. You can even adjust your thermostat, turn on the coffee pot, and listen for intruders—all from your cell phone across town.

How It Works

X-10 products require little or no wiring. In most cases, you simply plug devices into existing wall outlets or screw them into light sockets, and then turn a couple of dials. A typical X-10 system includes a variety of controllers, modules, and switches. Each device transmits or receives high-frequency signals, which travel along your home's electrical wiring or through the air as radio waves. Modules receive these signals from controllers to operate lights, alarms, and appliances.

Most controllers, modules, and switches must be programmed with house and unit codes. Each component's face features a red dial labeled *A* through *P* and a black dial labeled 1 through 16. The red dial sets the house code, which identifies the devices as part of the same system and prevents accidental operation by a neighbor with an X-10-compatible setup. The black dial controls the unit code, which makes appliances work together or on their own. Set a group of lights to the same unit code, and they'll switch on or off simultaneously. X-10-compatible systems provide 256 possible house/unit-code combinations.

A Basic X-10 Installation

First, go through each room in the house and decide which doors, windows, and areas you want protected from intruders. Also decide which lights and appliances you want the system to control. Once you install a basic system, you can add components later as your security needs grow. After choosing components, use a screwdriver to set house and unit codes.

The plug 'n power supervised security-console dialer with hand console controls the system. The dialer operates up to 16 groups of lights, appliances, and alarm sensors. When the alarm is tripped, the dialer also phones up to four numbers and plays a recorded message. The pocket-sized hand console controls up to four lights and appliances.

Install the dialer close to an electrical outlet and phone jack, but beyond an intruder's easy reach—a nightstand in the master bedroom usually works best. After flipping the dialer's mode switch to Install, attach a 9-V battery as a backup in case of power failure. Plug the unit in, raise its antenna, and push the unit's earphone into its jack. Then, run the phone cord from the unit to the phone jack.

Next, program four emergency numbers into the dialer the same way you store speed-dial numbers on a telephone. Finish off by recording a 13-second message, such as, "A burglary may be in progress at John Smith's house. The burglar alarm was tripped. Press zero to listen in, and call the police if necessary."

Before testing the alarm, call the first person on the programmed recording list and explain the system's operation. Set the dialer's mode switch to Run 2, and then press Arm on the hand-held remote to arm the system. Then, press the remote's Panic button to trip the alarm. Be sure you can hear the dialer's built-in siren throughout the house or office.

Once the alarm goes off, the dialer contacts the first number. After the listener hears the message, they can press zero to shut off the alarm and listen for sounds of an intruder. When you finish the test, press Disarm on the remote, and then flip the dialer's switch back to Install.

Wall Outlets and Modules

Wall outlets control lamps or appliances plugged into them. Wall switches operate indoor and outdoor overhead lights with an X-10 controller. Be sure to turn off your home's electricity at the main circuit breaker before installing outlets and switches.

Remove the existing wall outlet's cover plate and pull the outlet out of the electrical box. Then, disconnect the wires running from the box to the outlet. When connecting the new outlet, simply match the wire colors. If no ground wire extends from the electrical box, connect the module's green wire to the box. Push the module into the box, and then install the new cover plate. Then, follow the same steps to install a remote switch.

You can install the Anywhere Wall Switch on any flat surface in the house. It requires four AAA batteries and operates four light fixtures without running wires. Use Velcro fastening tape or screws to mount the module. To operate lamps, a coffee maker, or other small appliance, plug the Two-Prong Polarized Lamp or Appliance Switch into a wall outlet. You can control any appliance through the system once it's plugged into the switch.

Troubleshooting

If your X-10-based system works poorly even after proper installation, there are two likely causes: lack of phase coupling and power-line noise, also called *interference*. Power enters your home as 220/240-V service from two hot wires, called *phase A* and *phase B*, and your outlets are divided between them. If you plug a transmitter into a phase A outlet and a receiver into a phase B outlet, the transmitter may have to send its signal to the outdoor transformer before it reaches the receiver. By the time the signal arrives, it may be too weak for the unit to work properly. Check the circuit box diagram to find each outlet's phase, and to make sure controllers and switches match. An electrician can install a phase coupler at the circuit breaker to bridge the phases.

Fluorescent lights, computers, televisions, and other appliances can produce noise that interferes with X-10 signals. If your system has a problem, unplug appliances one at a time to find the culprit. Eliminate the noise by keeping the offending appliance unplugged or installing noise filters. If fluorescent lights are the only source of line noise, try replacing their ballasts because some versions produce less noise.

Chapter Quiz

1. "Home automation" is a generic term that refers to any automated technology used in homes.

    A. True   **B.** False

2. If your X-10-based system works poorly even after proper installation, there are two likely causes: lack of phase coupling and power-line noise, also called *interference*.

    A. True   **B.** False

3. Florescent lights, computers, televisions, and other appliances can produce noise that interferes with X-10 signals.

    A. True   **B.** False

4. A Smart House uses a system controller instead of a fuse panel, and Smart blocks instead of standard electrical outlets.

    A. True   **B.** False

5. A big difference between standard electrical outlets and Smart blocks is that electricity is always present in the standard outlets.

    A. True   **B.** False

6. If you were to stick a pin into a Smart block, you wouldn't get shocked because no electricity would be present.

    A. True   **B.** False

7. Once you install a basic X-10 system, you can't add components later as your security needs grow.

    A. True   **B.** False

8. If your customer's refrigerator door has been left open, the Smart House could signal their television set to show a picture of a refrigerator in the corner of the screen until the door is closed.

    A. True   **B.** False

9. With an X-10 system wall outlets control lamps or appliances plugged into them.

    A. True   **B.** False

10. With an automated security system in place, you can operate house lights and appliances or trigger the alarm siren with a hand-held remote control device.

    A. True   **B.** False

Chapter 14

FIRE PROTECTION

Vying with Canada during the past two decades, the United States continues to have one of the worst fire death records among industrialized countries. Most fire deaths in North America occur in homes and could have been avoided if the victims had taken simple precautions.

Many people in the United States and Canada don't take fire safety seriously. During school fire drills, for instance, teachers and students stand outside talking and giggling. We tend to feel sympathy for a person who experiences a home fire. In Great Britain and other countries, fire victims are penalized for their carelessness. Perhaps the contrast in attitudes has something to do with the difference in fire death rates.

This chapter looks at how home fires occur, and how people can avoid and survive them. You need this information to help your customers best choose and use fire safety products. You also need this information if you want to be known as a security consultant, instead of just a locksmith, safe technician, or alarm systems technician.

Causes and Cures

According to the U.S. Fire Administration, most home fires can be traced to smoking, cooking, heating equipment, and electrical appliances. More civilians die in fires related to in-house smoking than any other type of fire. Over 90 percent of fire deaths each year are the result of someone falling asleep or passing out while holding a lighted cigarette or while a lighted cigarette was burning out on a nearby furniture surface or in a wastebasket. Mattresses, stuffed chairs, and couches often trap burning ashes for long periods of time while releasing poisonous gases. Many people are killed by the gases rather than by heat.

The best way to avoid smoking-related fires is not to smoke in the home. If your customer is a smoker or allows other people to smoke in their home, they should be sure that sturdy, deep ashtrays are in every room in which people smoke. They should also always douse butts with water before dumping them in the trash, and check under and behind cushions for smoldering butts before leaving home or going to bed. No one should smoke when they're drowsy or while they're in bed.

The kitchen, where people work with fire most frequently, is the leading room of origin for home fires. Here are some simple things anyone can do to virtually eliminate the risk of ever having a major kitchen fire:

1. Keep the stove burners, oven, and broiler clean and free of grease.

2. When cooking on the stovetop, never leave it unattended.

3. Turn the handles of pots and pans away from the edges of the stove while cooking.

4. When cooking, wear short sleeves or keep sleeves rolled up (to avoid dragging them near the flames).

5. Make sure no towels, paper, food wrappings or containers, or other flammable items are close to the stove.

6. Don't use towels as pot holders (towels ignite too easily).

7. Never store flammable liquids in the kitchen.

It's possible to still have a small grease fire occasionally while cooking. (Be especially careful when frying foods.) Be prepared to respond immediately to such a fire. Respond to a small grease fire on the stove by turning off all burners on the stove and quickly covering the burning pan with a large metal lid. If no metal lid is at hand, pour a large quantity of flour onto the burning area to smother the flames while you get a cookie sheet to place of top of the pan to seal off the oxygen—

or use your fire extinguisher to put the fire out. Don't pick up the pan and carry it to the sink. You may burn yourself, spill burning grease, or drop the pan and start a fire on the floor.

Although more fires start in kitchens than in any other rooms, cooking isn't the main culprit. The number one cause of home fires is heating equipment. Nearly one-fourth of home fires involve space heaters, fireplaces, or wood stoves.

To avoid a heating equipment fire:

- Make sure any heating equipment you buy is tested and approved by an independent testing laboratory (such as Underwriters Laboratories).

- Be sure to follow the manufacturers' instructions when using the equipment.

- Never leave flammable materials near heating equipment.

- If you use a space heater, always keep it at least 36 inches away from anything combustible, including wallpaper, bedding, and clothing.

- At the start of each heating season, make sure the heating system is in good working order. Check standing heaters for fraying or splitting wires and for overheating. If any problems are noticeable, have all necessary repairs done by a professional.

During a typical year in the United States, home appliances and wiring problems account for about 100,000 fires and over $760 million in property loses.

Many fires could have been prevented if someone simply had noticed a frayed or cracked electrical cord and had it replaced.

You may think most of the preceding fire safety suggestions are so obvious that they don't need to be stated. They are "obvious," but everyday fires occur because someone failed to take one of those simple precautions. In addition to following those suggestions, you should have a few safety products, such as smoke detectors and fire extinguishers.

Smoke Detectors

A working smoke detector is the single most-important home safety device. About 80 percent of all fire deaths occur in homes not equipped with enough working smoke detectors. Most fatal fires happen between midnight and 4 a.m., when residents are asleep. Without a working smoke detector, people may not wake up during a fire because smoke contains poisonous gases that can put people into a deeper sleep.

The vast majority of homes in the United States have at least one smoke detector installed, but most of the detectors don't work because their batteries are dead or missing. Simply having a smoke detector isn't enough. It has to be in working order to help people stay safe.

There are two basic types of smoke detectors: ionization detectors and photoelectric detectors. They work on different principles, but either type is fine for most homes. Considering that many models sell for less that $10, it's foolish not to have several working smoke detectors in a home or office.

Smoke detectors should be installed on every level of a building, including the basement. A detector should be placed directly outside each sleeping room. The best location is 6 inches away from air vents and about 6 inches away from walls and corners.

Test all smoke detectors once a month to make sure they're in good working order. If they're battery-operated, replace the batteries as needed—usually about twice a year. Some models sound an audible alert when the battery is running low. Don't make the mistake of removing smoke detectors' batteries to use them for operating something else.

The Kidsmart Vocal Smoke Alarm

Traditional smoke detectors do not reliably awaken sleeping children. This is not so because the detectors aren't loud enough, but rather because our brains respond better to a familiar sound when we are sleeping than to the shrill tone of a conventional alarm. This recently discovered problem has been documented by media stations across the United States.

And, the solution—a personally recorded "familiar voice message"—has been studied by respected institutions from around the globe, including the Victoria University Sleep Laboratory of Melbourne, Australia, the world's foremost authority on sleeping and waking behaviors. In those tests, Dr. Dorothy Bruck discovered that 100 percent of all children tested with a "familiar voice" awoke within seconds.

Additionally, tests have either been conducted or are currently ongoing by the following institutions: Consumer Product Safety Commission, U.S. Naval Academy Fire Department, University of Georgia, and others.

Fire Extinguishers

A fire extinguisher can offer good protection if the right model is used, and if the user knows how and when to use it. If the wrong type is used, it can make the fire spread. There are several types of fire extinguishers, and each type is designed to extinguisher fires from particular sources. The main types of fire extinguishers are:

- *Class A*—for wood, paper, plastic, and clothing fires

- *Class B*—for grease, gasoline, petroleum oil, and other flammable liquids fires

- *Class C*—for electrical equipment and wiring fires

For most homes and offices, a good idea is to buy a class ABC fire extinguisher, because it's useful for a wide range of types of fires.

Buy a fire extinguisher that everyone in the home or office will find easy to use. A fire extinguisher won't be much good if no one is strong enough to lift it. Look for a model with a pressure-gauge dial. Then, your customers will know at a glance when the pressure is low and the extinguisher needs to be refilled.

When you sell a fire extinguisher, advise your customer to read the instructions carefully. That will help them be ready to use it correctly and without hesitation at any time. In most cases, they should stand at least 8 feet away from the fire, remove a pin from the extinguisher, aim the nozzle at the base of the fire, and squeeze the trigger while sweeping the nozzle back and forth at the base of the fire until they're sure the fire is out. An easy way to remember how to use the fire extinguisher is to remember the acronym PASS, which stands for pull, aim, squeeze, and sweep.

Emphasize to your customer that owning a fire extinguisher doesn't make them a firefighter, and that visible flames are only one lethal element of a fire. Unless it's a small fire that can be quickly put out, the customer should call the fire department. A fire extinguisher is only for putting out a small fire, but many small fires can spread quickly, and then become uncontrollable and life-threatening.

Escape Ladders

If your customer lives in a multiple-story home, they should plan a way to escape safely from windows located above the ground floor. One option is to install rope-ladder hooks outside each upper-floor bedroom and keep a rope ladder in each of the bedroom closets. Another option is to use a fixed ladder, such as the Redi-Exit.

The *Redi-Exit* is a unique ladder disguised as a downspout when not being used. Its shape discourages people from trying to use it to gain entry into a home. From an upper-floor window, you can open the Redi-Exit by striking down on a release knob. The unit can be installed on a new or an existing home.

Fire Sprinkler Systems

Studies by the U.S. Fire Administration indicate the installation of quick-response fire sprinkler systems in homes could save thousands of lives, prevent a large portion of fire-related injuries, and eliminate hundreds of millions of dollars in property losses each year. Sprinklers are the most reliable and effective fire protection devices known because they operate immediately and don't rely on the presence or actions of people in the building. Residential sprinklers have been used by businesses for over a century, but most homeowners haven't considered installing them because they are misinformed about sprinklers and misunderstand their use.

One misconception about residential sprinklers is that all of them are activated at once, dousing the entire house. In reality, only the sprinkler directly over the fire goes off because each sprinkler head is designed to react individually to the temperature in that particular room. A fire in a kitchen, for example, won't activate a sprinkler head in a bedroom.

Another misconception is that fire sprinklers are prohibitively expensive. A home sprinkler system can cost less than 1 percent of the cost of a new home—about $1.50 per square foot. The additional cost may be minimal when spread over the life of a mortgage. You may find a home sprinkler system virtually pays for itself in homeowner's insurance savings. Some insurers give up to 15 percent premium discounts for homes with sprinkler systems.

If your customer can't see their way clear to installing a full-blown sprinkler system, suggest one that protects one of your most vulnerable areas—the kitchen stove. The Guardian is the first automatic range-top fire extinguisher available for home use. *It was developed for U.S. military use after a 1984 study identified cooking-grease fires as the number one cause of fire damage and injuries in military-base housing.* The patented system uses specially calibrated heat detectors to trigger the release of a fire-extinguishing chemical.

When the chemical is released, the system automatically shuts off the stove. In laboratory tests, The Guardian has been found to detect and extinguish stove-top fires within seconds—but not to activate under normal cooking conditions. You can install it so it also activates an alarm inside a home.

The Guardian is UL-listed and combines a fire-detection assembly and a chemical distribution system into a single automatic unit. The fire-detection system can be installed neatly under any standard range-top hood. Cables connect it to the extinguisher assembly, which is housed in the cabinetry above the stove top. A pressurized container stores a fire-extinguishing liquid that is distributed through stainless steel piping to the underhood nozzles.

Here's how The Guardian responds when a stove-top fire starts:

1. Extreme heat from the stove-top fire causes any of four fusible links in the underhood detection assembly to separate, releasing tension on a cable.

2. When the cable tension is released, a tension spring automatically opens the extinguisher valve, discharging the liquid extinguishing mixture through the piping.

3. The mixture flows through two nozzles positioned directly above the stove-top burners, and a controlled discharge smothers the fire and guards against another fire starting.

4. While the extinguishing mixture is being released, a microswitch activates a switch that shuts off the gas or electric fuel source.

Surviving a Home Fire

To ensure that your customer and their family will be able to get out alive during a fire, they need to plan ahead. All members of the household, including small children, should help to develop an escape plan and regularly practice using it. It isn't enough for them just to say what they would do in case of a fire. They may have only seconds to get out, and the smoke could be so thick and black, they won't be able to see where they're going. Only through practice will they be able to react quickly and do almost routinely what they need to do to survive.

Make sure all potential escape routes are readily accessible. Check that windows aren't painted shut. Remove furniture blocking exit doors. Adjust locks that are too high for children to reach. And so on. Let your customer know they should take care of *any* obstacles immediately.

Tell your customer to establish a meeting place outside and not too close to their home (a spot near a designated tree or on a neighbor's porch, for instance). Agree that all members of the household will go there and wait together for the fire department. Everyone should know how to call for help—either at a neighbor's home or by using a fire box.

Emphasize the importance of not going back into a burning home, even if someone is unaccounted for. If someone goes back into the home, they will not only being endangering themselves, but also anyone inside. Fire grows quickly, and it rushes to wherever there is oxygen. As windows and doors are opened in a burning home, it makes the smoke and flames spread faster. Staying outside and waiting for the firefighters is better. Firefighters will arrive quickly, and they will have the equipment and skills to rescue anyone left inside.

Here are some key actions everyone should remember. If someone encounters smoke on the way out of the building, they should use an alternative exit. If someone must escape through smoke, they should crawl along the floor, under the smoke, where the air is cooler and cleaner.

If a person's clothing catches on fire, they should stop, drop to the ground, and roll to extinguish the flames.

If someone is in a bedroom and hears a smoke detector but doesn't see smoke, they should leave quickly through a bedroom window, if possible. If the room is too high off the ground or the person can't get out of the window safely, the person should feel the door from the bottom up to find out if it's warm. The person shouldn't touch the door knob, because it may be hot. If the door feels warm, they should *not* open it. If the door is cool, the person should place their shoulder against the door and open it slowly. If no flames can be seen and an exit is near, the person should crawl to safety. Once they are out of the building, the person should call the fire department immediately and not go back into the building for any reason. If everyone in the home has practiced what to do in case of a fire, your customer and their loved ones will know what to do to stay safe while waiting for the firefighters.

If the bedroom door is hot and someone can't safely climb out of a window, they should stuff rags or rolled-up clothes under and around the door, and in every gap or opening that may allow smoke to enter the room. If someone can't climb out of a window safely, they should hang a rag or a piece of clothing out of it. This will let firefighters know where the person is located.

What to Do after a Fire

If you have a home fire, take these actions as quickly as you can afterward:

1.   Immediately call your insurance company or the insurer's agent, and then call your mortgage company.

2.   Don't let anyone into your home without first seeing identification. Criminals may try to take advantage of your vulnerable situation.

3. Make sure all your utilities are turned off. If you're in a cold climate and you expect your house will be empty for a long time, drain the water lines.

4. Protect all undamaged property to avoid further damage.

5. Don't clean up until after your insurance company has inspected the damage.

6. Make a list of all your damaged property. If possible, include the model numbers, serial numbers, dates and places of purchase, and purchase prices. The more details you have about your property, the better off you'll be when dealing with your insurance company.

7. If your home is too damaged to live in and you need temporary shelter, call your insurance company, the local Red Cross, or the Salvation Army for help. Other possible sources of help include churches and synagogues, and civic groups, such as the Lions Clubs International and the Rotarians.

8. Keep all receipts for additional living expenses and loss-of-use claims.

Be wary of uninvited insurance adjusters who may contact you after hearing a report of the fire. If you have a complete inventory of your property and it's readily accessible, an insurance adjuster probably can't do any more for you than you can do for yourself.

Chapter Quiz

1. Most home fires can be traced to smoking, cooking, heating equipment, and electrical appliances.

   A. True   **B.** False

2. Once someone gets out of a burning building, the person should not go back in unless a family member is unaccounted for.

   A. True   **B.** False

3. The bedroom is the leading room of origin for home fires.

   A. True   **B.** False

4. The best way to avoid smoking-related fires is to smoke cigars.

   A. True   **B.** False

5. To help avoid kitchen fires the stove burners, oven, and broiler should be clean and free of grease.

   A. True   **B.** False

6. Most fire deaths are the result of someone falling asleep or passing out while holding a lighted cigarette, or while a lighted cigarette was burning out on a nearby furniture surface or in a wastebasket.

   A. True   **B.** False

7. You should respond to a small grease fire on the stove by turning off all burners of the stove and quickly covering the burning pan with a large metal lid, if it's safe to do so.

   A. True   **B.** False

8. The number one cause of home fires is heating equipment.

   A. True   **B.** False

9. More fires start in kitchens than in any other room.

   A. True   **B.** False

10. A space heater should always be kept at least 5 inches away from anything combustible, including wallpaper, bedding, and clothing.

    A. True   **B.** False

11. A working smoke detector is the single most-important home fire safety device.

    A. True   **B.** False

12. Most fire deaths occur in homes not equipped with enough working smoke detectors.

    A. True   **B.** False

13. Smoke detectors should be installed on every level of a building, including the basement.

    A. True   **B.** False

14. A class ABC fire extinguisher is best for most homes and offices.

    A. True   **B.** False

15. In most cases, a person using a fire extinguisher should stand at least 8 feet away from the fire, remove the pin from the extinguisher, aim the nozzle at the base of the fire, and squeeze the trigger while sweeping the nozzle back and forth at the base of the fire until the fire is out.

    A. True   **B.** False

16. Residential sprinklers all activate at once during a house fire.

    A. True   **B.** False

17. Class *A* fire extinguishers are for wood, paper, plastic, and clothing fires.

    A. True   **B.** False

18. Class *B* fire extinguishers are for grease, gasoline, petroleum oil, and fires caused by other flammable liquids.

    A. True   **B.** False

19. Class *C* fire extinguishers are for electrical equipment and wiring fires.

    A. True   **B.** False

20. The Redi-Exit is a ladder disguised as a downspout when it's not being used.

    A. True   **B.** False

21. If a person's clothing catches on fire, they should stop, drop to the ground, and roll to extinguish the flames.

    A. True   **B.** False

22. After a home fire, the homeowner should immediately call their insurance company or the insurer's agent, and then call their mortgage company.

    A. True   **B.** False

Chapter 15

SAFETY AND SECURITY LIGHTING

Lighting is not only a low-cost form of security, but it also can help to prevent accidents, create moods, and enhance the beauty of any home. This chapter shows how you can make the best use of lighting inside and outside your home.

A dark house is an invitation to crime and creates a high risk for accidents. When you approach your home late at night, you need to be able to walk to your entrance without tripping over something—or *someone*—in your path. Once you're inside, you need to be able to move from room to room safely. Your home should be *well lighted* on the inside, in the areas directly outside the doors, and throughout the yard. Well lighted doesn't necessarily mean a lot of light; it means having the light sources and controls in the right places.

Light Sources

Our most common light source is the sun, which we cannot control. We have artificial light sources available for use at night and in some indoor locations during the day. Important differences among artificial sources include color, softness, brightness, energy efficiency, and initial cost.

Light sources you might consider for home use include standard incandescent, halogen, fluorescent, and high-intensity discharge (HID) lighting. The HID family of lighting includes low-pressure sodium, high-pressure sodium, mercury-vapor, and metal halide lights.

An *incandescent* light source relies on heat to produce light. The standard bulbs used in most homes are incandescent (lighting designers call them *A-lamps*). They have a metal filament that is heated by electricity. Standard incandescent bulbs are inexpensive, readily available, and suitable for most home fixtures. They light almost immediately at the flip of a light switch. However, using heat to produce light isn't energy efficient; in the long run, incandescent lighting can be more costly than other sources that require special fixtures.

Halogen, a special type of incandescent source, is slightly more energy efficient than standard incandescent lighting. A halogen bulb uses a tungsten filament and is filled with a halogen gas.

Fluorescent lighting uses electric current to make a specially shaped (usually tubular) bulb glow. The bulbs come in various lengths, from 5 inches to about 96 inches, and they require special fixtures. You might not want to use fluorescent lighting with certain types of electronic security systems because it can interfere with radio reception. And you wouldn't ordinarily use fluorescent lighting outdoors in cold climates: it's very temperature-sensitive.

For outdoor lighting, you might use high-intensity discharge (HID) sources, which are energy-efficient and cost little to run for long periods of time. Like fluorescent lighting, HID sources require special fixtures and can be expensive initially. Another potential problem with HID sources is they can take a long time to produce light after you turn them on. Startup time can be unimportant if you use a light controller to activate the lights automatically when necessary.

Light Controllers

Timers are among the most popular types of controllers for indoor and outdoor lighting. The newest timers can do much more than just turn lights on and off at preset times. Programmable 24-hour wall-switch timers, for example, can turn your lights off and on randomly throughout the night and early morning. This feature is useful because many burglars will watch a home to see whether the lights come on at exactly the same time each night—an indication the home is empty and a timer was pre-set. Another feature of some new timers is built-in protection against memory loss. After a power failure, such timers "remember" how you programmed them. Some models adjust themselves automatically to take into account daylight saving time changes. Most timers used by homeowners cost between $10 and $40.

> ## Installing Motion-Activated Outdoor Lighting
>
> A motion-activated light can be installed virtually anywhere indoors or outdoors—on the side of home, on a porch, in a garage—wherever it's needed. Many of these lights are simple two-wire installations in which the hardest part of the process is the proper positioning of the lighting unit. Models are available in various colors and styles to match any decor.

Another low-cost way to control lights is with sound or motion sensors. You can buy one of these sensors and connect it to, say, a table lamp in your living room. When you (or someone else) walk into your living room at night, the light will come on automatically.

Some *floodlights* come with a built-in motion sensor. If you install floodlights outside at strategic places, they will warn you of nighttime visitors. You might install one facing toward your driveway, for instance, so it lights the area when a car pulls in. Floodlights generally sell for less than $50.

Preventing Accidents

To prevent accidents at night, you need to be able to see potential hazards. When walking down a flight of stairs, for instance, it's important to be able to see whether any objects are in your way. In too many homes, people have to stumble through dark areas or grope for a light switch, or a series of light switches, before reaching the bathroom or kitchen.

Can you use motion-activated sensors to avoid this problem? You'd probably need many of them to cover every path you might take at night. Simpler and more convenient options are available.

One useful practice is to install night lights near your light switches, so you can reach them more easily. Night lights cost only a few dollars each, and they consume very little power. Another option is to use three-way switches. These let you turn a light on and off at more than one location, such as at the top and bottom of a flight of stairs.

The most convenient way to use lighting indoors is to tie all (or most) of the switches into an easily accessible master control panel. Then, you could turn on a specific group of lights simply by pushing a button. One button could turn on a pathway of lights from, say, your bedroom to the bathroom. By pushing another button, you could activate a pathway of lights from your bedroom to the kitchen. Some or all of your outdoor lights also could be tied into your master control system.

120-Volt Lighting

A *120-V lighting system* can provide brighter light than a low-voltage system. *The brighter light may be especially useful outside if you need to illuminate a large area.* Installing a 120-V system is more involved and the materials are more expensive than those used in a low-voltage system.

Before beginning the installation, familiarize yourself with your local electrical code, and obtain any required permits. You may have to draw up a plan and have it reviewed by your local building inspector before you're allowed to install 120-V lighting.

You need to decide what materials to use—receptacles, cables, switches, boxes, conduit, conduit fittings, wire connectors, and so on. Your local code may already have made some of these decisions for you. You may be required to use rigid metal conduit rather than PVC conduit, for instance, or you may be restricted to using only certain types of wire.

Chapter Quiz

1. A 120-V lighting system can provide brighter light than a low-voltage system.

   A. True   **B.** False

2. The most convenient way to use lighting indoors is to tie all (or most) of the switches into an easily accessible master control panel.

   A. True   **B.** False

3. Lighting not only improves security, but it also helps to prevent accidents, create moods, and enhance the beauty of a home.

   A. True   **B.** False

4. "Well lighted" means having bright lights.

   A. True   **B.** False

5. Important differences among artificial sources include color, softness, brightness, energy efficiency, and initial cost.

   A. True   **B.** False

6. The HID family of lighting includes low-pressure sodium, high-pressure sodium, mercury-vapor, and metal halide lights.

   A. True   **B.** False

7. An *incandescent* light source relies on heat to produce light.

   A. True   **B.** False

8. Standard incandescent bulbs are inexpensive, readily available, and suitable for most home fixtures.

   A. True   **B.** False

9. A halogen bulb uses a tungsten filament and is filled with halogen gas.

   A. True   **B.** False

10. Fluorescent lighting uses electric current to make a specially shaped (usually tubular) bulb glow.

    A. True   **B.** False

Chapter 16

CLOSED-CIRCUIT TELEVISION SYSTEMS

Locks and other physical security devices can be more effective when used in conjunction with a closed-circuit television (CCTV) system. Such a system can allow numerous areas—such as elevators, entrances, and exits; parking lots; lobbies; and cash-handling areas—to be monitored constantly. Such monitoring can deter crime and reduce a person's or company's security costs.

This chapter looks at the various components that make up a system and shows various ways in which such a system can be used in homes and businesses.

Basics

To begin selling and installing CCTV systems, you don't need a strong background in electricity or electronics. Although few locksmiths can handle the large multiplex systems used in airports, banks, and casinos, most CCTV systems used in homes, as well as in small offices, stores, and apartments, are easy to install. Many CCTV systems come preconfigured as a complete package.

However, it's still important for you, as a security professional, to have a basic understanding of each of the components you may be working with. This lets you be more helpful to your customers, which can mean more profits for you.

How CCTV Systems Work

A *CCTV system* transmits images to monitors that are connected to the system's camera. The system's basic components are a video camera and monitors connected to it by coaxial cable. This type of installation wouldn't be very useful for security purposes. You would have to prop the camera up in a room, point it to a fixed location you wanted to protect, and then go and stare at the monitor.

For security, you need a camera that works while you're not around, can be controlled from a remote location, and can be connected to a burglar alarm system. If you need to monitor more than one location, you also may want the option of monitoring both cameras at once. All these and many other features are possible with low-cost CCTV systems.

When you know what's available, you can choose the cameras, monitors, and optional components that will create a custom CCTV system within your customer's budget. If the system isn't too complex, you can probably install it yourself. Most of the CCTV systems used by homeowners and small businesses are easier to install than a hardwired alarm system.

Cameras

Two types of cameras are used commonly in CCTV systems: the tube camera—the older type, which is fast becomingly obsolete—and the closed-coupled-device (CCD) camera, which lasts longer and works better. CCD cameras cost a little more than tube cameras, but they have been steadily coming down in price, whereas the price for tube cameras has remained the same. With the demand for CCD cameras continuing to outpace the demand for tube cameras, many camera makers are discontinuing their line of tube cameras.

Cameras come in color and in black-and-white transmission models. A color camera requires maximum and constant light to be able to view a scene properly. It shouldn't be used outdoors or in any area that sometimes gets dark. A black-and-white camera is both more tolerant of all types of lighting conditions and less expensive.

The choice between color and black-and-white transmission is usually simple. In virtually every residential situation, black-and-white transmission is more cost-effective and much less troublesome. Color cameras are needed only in banks and at other surveillance sites where the cameras' videotapes may become evidence in a court case.

Cameras come in many sizes, described by their lens diameter. The three most common sizes are ⅓ inch, ½ inch, and ⅔ inch. The ⅔-inch camera covers more area and gives better resolution than the ⅓-inch camera. Generally, the larger the camera, the better the picture.

Monitors

In many ways, your choice of a monitor is as important as your choice of a camera. The quality of the picture you receive on your monitor depends on both choices. The camera and monitor work together, much like speakers and an amplifier in an audio system. If you have a great amplifier with poor speakers or great speakers with a poor amplifier, you'll get poor sound because the sound is filtered through both devices before you hear it. In a CCTV system, the picture is filtered through both the camera and the monitor before you see it.

For transmission of a color image, a monitor and a camera must be color equipment. If either is black-and-white, you'll receive a black-and-white picture. Monitors, described based on their screen diameter, range in size from about 4 inches to over 21 inches. The most common monitors for home use are the 9- and 12-inch sizes.

To save your customer money, you can suggest they buy a radio frequency (RF) modulator for their client's television to convert it to a monitor. Then, the customer will be able to view the camera's visual field just by turning the television to a particular channel (usually either Channel 3 or Channel 4).

Peripheral Devices

One of the most popular devices for CCTV systems is a *pan-and-tilt unit*, which gives a camera the capability to tilt up and down and to rotate up to 360 degrees, left to right or right to left. With a pan-and-tilt unit, someone can zero in on items (and people) within a wider camera range. By using a pan-and-tilt unit in a large installation, you can use only one camera, instead of several.

Pan-and-tilt units have long been used in airports, banks, and other large commercial installations. Because the units often cost over $1,000, they're rarely included in home CCTV systems.

Another complementary device you can recommend with a CCTV system is a sequential switcher. With one monitor, the *sequential switcher* will let someone receive pictures from several cameras. They can watch one camera field for a while, and then switch over to another.

If your customer wants a continuous record of what the camera sees, you can install a *video lapse recorder*, which will span up to 999 hours with individual photo frames on one standard VHS 120 videotape. If you want, the current time and date can be recorded automatically on each frame.

If your customer wants to record only unwanted persons who enter a particular area or room, you can use a camera with a built-in motion-detecting capability, which is connected to an alarm system. The alarm will be triggered when the camera begins taping.

Another option available with today's CCTV systems is the *dual-quad unit*, which gives a standard monitor the capability of showing as many as 16 pictures at one time from 16 separate cameras. Dual-quad units can cost anywhere from $1,000 to $15,000.

Installing a CCTV System

The specific installation methods that are best for you depend on the components you've chosen and how you want to use them. Many of the hardwiring methods for installing burglar alarms, detailed in Chapter 12, are useful for installing CCTV systems.

Most CCTV systems can be installed either independently or incorporated into a burglar alarm system. If you tie the CCTV system into a burglar alarm system and use a videotape recorder, you can set the CCTV system to begin recording automatically whenever the alarm is triggered. You also can have the system record sounds.

Your installation can be either overt or covert. Most homeowners use an overt system because they want would-be burglars to know they are being watched. A camera that's prominently connected to the side of a home certainly would act as a deterrent.

Some people consider cameras inside a home unattractive and threatening, but there are some advantages to keeping the cameras out of sight. Hidden cameras allow you to make a secret video-tape of a burglar.

To install a covert system, you need to buy small cameras specially designed for installation in the corner of a wall or in a wall cutout. They sell for between $100 and $200. Some covert cameras are disguised to look like clocks and other common objects. Their prices start at about $200.

You should seek legal advice before installing a covert system; in some jurisdictions, such systems are illegal. Some states consider secret audio taping a violation of wiretap laws.

Video Intercoms

A *video intercom system* is a CCTV system that lets a person talk to the people they see through the camera. With many models, you can choose to see and hear a person without the person knowing you're home. Like burglar alarms, video intercoms can give a home a high-level of security—and they're usually easier to install than burglar alarms.

In some video intercom systems, the cameras, monitors, intercoms, and peripherals are all separate components. Other systems have integrated components, such as a monitor with a built-in intercom or a camera with built-in peripherals. Because they have fewer components, integrated units usually are simpler to install. They also tend to take up less space and look nicer. The main problem with most integrated systems is they can't be expanded to add on sophisticated peripherals.

Some models are designed to incorporate a variety of peripherals. Aiphone's Video Sentry Pan Tilt, for example, includes an integrated camera, a monitor, an intercom, and a motorized pan-and-tilt unit. Its camera can scan 122 degrees horizontally and 76 degrees vertically—up to four times the area visible with a fixed camera. A button on the monitor unit lets your customer control the panning and tilting actions of the camera unit.

Separate components allow more flexibility during installation. You can mount the outside intercom where visitors can reach it easily, for instance, and place a separate camera where they can't see it. Separate cameras can be installed easily on gates, near swimming pools, and at other outside locations.

Monitors range in size (based on the diameter of the screen) from 4 to about 20 inches. Most integrated units have a built-in, 4-inch, black-and-white monitor. The 9- and 12-inch sizes are popular for monitors used with separate units.

Lighting Considerations

Different cameras need different amounts of light to view a scene properly. Whether it comes from the sun, from starlight, or from light bulbs, the amount of light needed is measured in *lux*. The fewer lux a camera needs, the more adaptable it is to night-time viewing. Most color cameras need a minimum illumination of 3 lux; black-and-white cameras usually need only 1 lux. Some security units require 0 lux because their cameras have built-in infrared diodes that produce the necessary amount of light.

Installation Tips

Is there a "best way" to install a video intercom? The installation method depends on the model your customer has chosen, where they want to use it, and how they want to use it. Most manufacturers will supply you with installation instructions, templates, and a wiring diagram. The following tips may make the job easier.

In a typical installation, you first need to decide where to mount the camera and the monitor units. Regardless of its minimum lux requirement, the camera should be placed where there's always enough light for someone to read a book page (a porch light or streetlight may provide enough light at night). With less light, your customer probably won't be able to see people well on the monitor screen. Don't position the camera so it will be subjected to direct, glaring sunlight. Another consideration is this: the place you choose should allow for easy wiring access to the indoor monitor.

The monitor unit can be placed in any location that's convenient for your customer (often in the room where they spend the most waking time) and near an electrical outlet. It's usually best to run your cable (you'll use coaxial two- or four-wire cable) before mounting the camera and monitor unit. In that way, if you have trouble getting the cable to the desired locations, you can choose other mounting locations without leaving unsightly screw holes.

Chapter Quiz

1. In a typical CCTV installation, you first need to decide where to mount the camera and the monitor units.

   A. True   **B.** False

2. A CCTV system can allow numerous areas—such as elevators, entrances, and exits; parking lots; lobbies; and cash-handling areas—to be monitored constantly.

   A. True   **B.** False

3. A color CCTV camera requires maximum and constant light to be able to view a scene properly.

   A. True   **B.** False

4. In most cases, black-and-white CCTV transmission is more cost-effective and much less troublesome than color.

   A. True   **B.** False

5. A pan-and-tilt unit gives a camera the capability to tilt up and down, and to rotate up to 360 degrees left to right or right to left.

   A. True   **B.** False

Chapter 17

HOME AND OFFICE SECURITY

Throughout this book, we've covered lots of information about safety and security systems, devices, and hardware. If you read all the chapters, you have the information necessary to think like a security consultant. This chapter discusses how to put everything together to make a home or business as safe as your customer or client wants it to be. Understand that no single security plan is best for everyone. Each home and business has unique strengths and vulnerabilities, and each building has different needs and limitations. The important limitation most people face is money. If money were no object, it would be easy to lay out a great security plan for any building.

With proper planning, your clients can be safe in their homes or businesses without spending more money than they can afford and without being too inconvenienced. Proper planning is based on the following considerations:

- How much money is the client willing to spend?

- How much risk is acceptable to your client?

- How much inconvenience is acceptable to your client?

- How much time is your client willing to spend on making their home or business more secure?

- How much of the work are your clients willing and able to do by themselves?

Before you can suggest security strategies for a home or business, you need to conduct a safety and security survey (or "vulnerability analysis"). This survey requires you to walk around the outside and the inside of the building, and take note of all potential problems.

Surveying a Home

The purposes of a safety and security survey are:

- To help you identify potential problems

- To assess how likely and how critical each risk is

- To determine cost-effective ways either to eliminate the risks or bring them to an acceptable level

The survey enables you to take precise and integrated security and safety measures.

A thorough survey involves not only inspecting the inside and the outside of a home or business, but also examining the safety and security equipment, as well as reviewing the safety and security procedures used by all employees or family members. The actions people take (or fail to take) are just as important as the equipment they may buy. What good are high-security deadbolts, for instance, if residents often leave the doors unlocked?

As you conduct your survey, keep the information in the preceding chapters in mind. You'll notice many potential safety and security risks (every home and business have some). Some of the risks will be simple to reduce or eliminate immediately. For others, you need to compare the risk to the cost of properly dealing with them. There's no mathematical formula to fall back on. You need to make subjective decisions, based on what you know about the household or business.

When surveying a building, it's best to start outside. Walk around the building and stand at the vantage points that passersby are likely to have. Many burglars will target a home or business because it's especially noticeable while driving or walking past it. When you look at the building from the street, note any feature that might make someone think the place is easy to break into or that it may have a lot of valuables inside.

Remember, burglars prefer to work in secrecy. They like heavy shrubbery or large trees that block or crowd an entrance, and they like buildings that aren't well lighted at night. Other things that may attract burglars' attention include expensive items that can be seen through windows, a ladder near the building, and notes tacked on the doors.

As you walk around the building, note anything that might help discourage burglars. Can a "Beware of the Dog" sign or your security system sticker be seen in the window? Walk to each entrance and consider what burglars might like and dislike about it. Is the entrance well lighted? Can neighbors see someone who's at the entrance? Is a video camera pointing at the entrance? Does the window or door appear to be hard to break into?

After surveying the outside of the home, go inside and carefully examine each exterior door, window, and other opening. Consider whether each one is secure, but allows occupants to get out quickly. Check for the presence of fire safety devices. Are there enough smoke detectors and fire extinguishers? Are they in working order? Are they in the best locations?

Take an honest look at the safety and security measures the occupants already have in place. What habits or practices would be helpful to change?

Home Safety and Security Checklist

Because every home and business is unique, no safety and security survey checklist can be comprehensive enough to cover all of every building's important factors. But, the following checklist can help guide you during your survey. Keep a notepad handy to write down details or remedies for potential problems.

Surveying an Apartment

In many ways, surveying an apartment is like surveying a house. The difference is you have to be concerned, not only about the actions of the household, but also about those of the landlord, the apartment managers, and the other tenants. The less security-conscious neighbors and others are, the more at-risk everyone will be. No matter how much one tenant does to avoid causing a fire, for example, a careless neighbor may cause one. Likewise, if neighbors don't care about crime prevention, all of the apartment building or complex will be more attractive to burglars.

As you walk around the outside of an apartment, notice everything that would-be burglars might notice. Will they see tenants' "crime watch" signs? Will they see that all the apartments have door viewers and deadbolt locks? Burglars hate a lot of door viewers because they never know when someone might be watching them.

After surveying the outside, walk through the apartment and look at each door, window, and other opening. If you notice major safety or security problems, point them out to your customer or client. You might also want to suggest little things the landlord can do to make the apartment more secure.

High-Rise Apartments

High-rise apartments have special security concerns that don't apply to apartments with fewer floors. In a high-rise, more people have keys to the building, which means more people can carelessly allow unauthorized persons to enter.

The physical structure of a high-rise often provides many places for criminals to lie in wait for victims or to break into apartments unnoticed. Many high-rise buildings aren't designed to allow people to escape quickly during a fire.

The safest apartments have only one entrance for tenants to use, and that entrance is guarded 24 hours a day by a doorman. An apartment that doesn't have a doorman should have a video intercom system outside the building. Video intercoms are better than audio intercoms because they let you see and hear who's at the door before you buzz the person in.

Home and Office Safety and Security Checklist

As you conduct your survey, note each potential problem of concern to you.

Home Exterior

_____ Shrubbery (Shouldn't be high enough for a burglar to hide behind—or too near windows or doors.)

_____ Trees (Shouldn't be positioned so a burglar can use them to climb into a window.)

_____ House numbers. (Should be clearly visible from the street.)

_____ Entrance visibility (Should allow all entrances to be seen clearly from the street or other public area.)

_____ Lighting (Should be near the garage or other parking area.)

_____ Ladders (Shouldn't be in the yard in clear view.)

_____ Mailbox (Should be locked or otherwise secured, and should either show no name, or a first initial and the last name.)

_____ Windows (Should be secured against being forced open, but should allow for an easy emergency exit.)

_____ Window air conditioners (Should be bolted down or otherwise protected from being removed.)

_____ Fire escapes. (Should allow for easy emergency escape, but should not allow for unauthorized entry.)

Exterior Doors and Locks

(Included here are doors connecting a garage to a home.)

_____ Door material (Should be solid hardwood, fiberglass, PVC plastic, or metal.)

_____ Door frames (Should allow doors to fit snugly.)

_____ Door glazing (Shouldn't allow someone to gain entry by breaking it and reaching in.)

_____ Door viewer (Every door without glazing should have a wide-angle door viewer or other device to see visitors.)

_____ Hinges (Should be either inside the door or protected from outside removal.)

_____ Stop moulding (Should be one piece or protected from removal.)

_____ Deadbolts (Should be single cylinder, with a free-spinning cylinder guard, and a bolt with a 1-inch throw and hardened insert.)

_____ Door openings, including mail slots, pet entrances, and other access areas (Shouldn't allow a person to gain access through them.)

_____ Sliding glass doors (Should have the movable panel mounted on the interior side, and a bar or other obstruction in the track.)

Inside the Home or Office

____ Fire extinguishers (Should be in working order and mounted in easily accessible locations.)

____ Smoke detectors (Should be in working order and installed on every level of the building.)

____ Rope ladders (Should be easily accessible to rooms above the ground floor.)

____ Flashlights (Should be in good working order and easily accessible.)

____ First aid kit (Should contain fresh bandages, wound dressings, burn ointment, aspirin, and rubber gloves.)

____ Telephone (Should be programmed to quickly dial the police and fire departments. Otherwise, keep the phone numbers close to the phone.)

____ Intruder alarm (Should be in good working order and adequately protected from vandalism, and they should have adequate backup power.)

____ Safes. (Should be installed so they can't be seen by visitors.)

Chapter Quiz

1. A sliding glass door should have a bar or other obstruction in the track.

   A. True   **B.** False

2. A deadbolt lock should be single cylinder with a free-spinning cylinder guard, and a bolt with a 1-inch throw and hardened insert.

   A. True   **B.** False

3. Every exterior door without glazing should have a wide-angle door viewer or other device to see visitors.

   A. True   **B.** False

4. Window air conditioners should be bolted down or otherwise protected from being removed.

   A. True   **B.** False

5. House numbers should *not* be clearly visible from the street.

   A. True   **B.** False

Chapter 18

COMPUTER SECURITY

Whether a computer runs Microsoft Windows, Apple's Mac OS, Linux, or something else, the security issues are similar and will remain so, even as new versions of the system are released. As a security professional, you need to understand the security-related problems, so you can help your clients—as well as yourself.

Home and business computers are popular targets for computer hackers because they want the information stored in them. Hackers and crackers look for passwords, credit-card numbers, bank account information, and anything else they can find. By stealing that information, they can use other people's money to buy themselves goods and services.

And hackers aren't only after money-related information. They also want to use your computer's hard-disk space, your processor, and your Internet connection. The intruder uses those resources to attack other computers on the Internet. The more computers an intruder uses, the harder it is for law enforcement to determine where the attack is originating. The intruder must be found before they can be stopped and prosecuted.

Intruders pay special attention to home and office computers because such computers aren't very secure and they're easy to break into. Because the computers often use high-speed Internet connections that are always turned on, intruders can quickly find, and then attack, home or office computers. They also attack computers connected to the Internet through dial-up connections, but hackers favor computers connected with high-speed cable and DSL modems. Regardless of how a computer is connected to the Internet, a computer is susceptible to hacker attacks.

Installing and Using an Antivirus Program Checklist

To make sure an effective antivirus program is installed, ask the following questions:

1. Do you have an antivirus program installed?
2. Do you check frequently for viruses before sending or receiving e-mail?
3. Do you check for new virus signatures daily?
4. Is your antivirus program configured to check every file on your computer (CD-ROMS, floppy disks, e-mail, and the Web)?
5. Do you have heuristic tests enabled?

A key to Internet security begins with a properly configured Internet *firewall*—software or hardware that helps to screen out hackers, viruses, and worms that try to reach your computer over the Internet. (Both Windows XP Home Edition and Windows XP Professional with Service Pack 2 (SP2) have a firewall already built in and active.) If you have Microsoft Windows XP (SP2) running on your computer, you can check to see if your firewall is turned on through the Windows Security System. Just click Start, and then click Control Panel. Next, click Security Center, and then click Windows Firewall. Versions of Windows before Windows XP did not come with a built-in firewall. If you have a different version of Windows, such as Windows 2000, Windows Millennium Edition (ME), or Windows 98, get a hardware or software firewall from another company and install it.

If you use Windows XP, but you want different features in a firewall, you can use a hardware firewall or a software firewall from another company. Many wireless access points and broadband routers for home networking have built-in hardware firewalls that provide good protection for most home and small-office networks. Software firewalls are a good choice for single computers, and they work well with Windows 98, Windows ME, and Windows 2000.

If you aren't sure which version of Windows you have, click Start, and then click Run. In the Run dialog box, type **winver**, and then click OK. The dialog box that appears will tell you which version of Windows software you're running.

If you have two or more computers in your home or office network, you need to protect each computer in the network. Enabling the Internet connection firewall on each connection can help to prevent the spread of a virus from one computer to the other(s) in your network if one of your computers becomes infected with a virus. If a virus is attached to an e-mail message, however, the firewall won't block it, and it can infect your computer.

You only need to run one firewall per computer. Running multiple software firewalls isn't necessary for typical home computers, home networking, and small-business networking scenarios. Using two firewalls on the same connection could cause problems with connectivity to the Internet. One firewall, whether it is the Windows XP Internet Connection Firewall or a different software firewall, can provide enough protection for your computer.

Antispyware

Spyware programs are small applications that can get installed on your computer without your knowledge. Such programs can get installed either by downloading innocent-looking software programs that include them or through ActiveX controls hidden within the source code of participating web sites or pop-up advertisements while you're surfing the Internet. These bundled programs and ActiveX controls can install a wide range of unwanted software onto a user's computer.

In addition to doing a detailed check of your browser history, spyware programs install a wide assortment of Dynamic Link Libraries (DLLs) and other executables files. They send a continuous data stream to the parent marketing company out from your computer and leave a back door open for hackers either to intercept your personal data or enter your computer. Spyware programs can install other programs directly onto your computer without your knowledge. They can send and receive *cookies*—small text files placed on your computer to keep tabs on you—to or from other spyware programs and invite them into your computer (even if you have cookies disabled). They can also bring Trojan horses into your system that perform a wide range of mischief, including changing your home page and downloading unwanted images and information.

Many spyware programs are independent executable files that are self-sufficient programs, which take on the authorization abilities of the user. They include automatic install and update capabilities, and they can report on any attempts to remove or modify them. These programs can hijack your home page; reset your browser favorites; reset your autosignature; disable or bypass your uninstall features; monitor your keystrokes on- or offline; scan files on your hard drive; access your word processor, e-mail, and chat programs; and change home pages. Many spyware programs can read, write, and delete files, and, in some instances, even reformat your hard drive. And, they do these things while sending a steady stream of information back to the advertising and marketing companies.

Most of these programs cannot be deleted from your system by normal methods and leave residual components hidden on your system to continue monitoring your online behavior and trying to reinstall themselves. Many people notice a big decrease in their computer's performance after installing spyware-infested programs—which use up your system resources.

New types of spyware are becoming more malicious: CoolWebSearch makes browsers useless by changing Internet Explorer (IE) settings and installing malicious applications; KeenValue collects information about users and sends advertisements to their systems; Perfect Keylogger logs keystrokes users enter, putting users' personal information at risk; and Marketscore redirects traffic from a host system to another that collects data before traffic reaches its final destination. Windows users should install and run an antispyware program, such as Microsoft AntiSpyware, Pest Control, Spyware Doctor, Spy Sweeper, and Spybot. AOL users with current software have built-in spyware protection.

Operating System Safeguards

Many operating systems (OSs) have good security features built into them, but such features are worthless if you don't know about them. One way hackers and crackers break into computers is by sending an e-mail with a virus in it. When you read the e-mail, the virus gets activated, creating an opening the intruder can use to access your computer. Other times, a hacker takes advantage of a flaw or weakness in one of your programs to gain access. When a virus gets into your computer, it may install new programs that let a hacker continue to use your computer, even after you plug the holes used to plant the virus in the first place. Such a "back door" often is cleverly disguised, so you won't recognize it.

Web browsers, such as IE, Mozilla Firefox, and Safari, are installed on most home and office computers. Because web browsers are used so often, it's important to configure them securely. Often the default settings for web browsers aren't set securely and, if web browsers aren't configured securely, hackers can easily gain control over your computer.

Compromised or malicious web sites often take advantage of vulnerabilities in web browsers either because the web browser's default settings are set to increase functionality, rather than provide security, or because new vulnerabilities are discovered after the web browser has been packaged and distributed by the manufacturer. It's important to know which features of your browser make it less secure. The Active X software feature, for example, has a long history of vulnerabilities. ActiveX is used on Microsoft IE and allows applications or parts of applications to be used by the web browser. It allows for extra functionality when web browsing, but it also creates security problems. Cross-site scripting (CSS or XSC) also creates vulnerabilities.

To increase your security, go to Internet Options, and set the security level to High. This is the safest way to browse the Internet, but you'll have less functionality and you may be unable to use some web sites. If you have trouble browsing a certain web site, send an e-mail to the Webmaster and ask them to design the site so you can browse it more securely.

If you want to browse a web site that you trust, set the security of your Trusted Site to Medium. Then, when you access a web site you trust that doesn't have malicious code, you can use ActiveX and Active Scripting. This will let you stay safe on most web sites and still have full functionality on those sites you trust.

If you're using IE, the privacy tab contains settings for cookies, which can contain any information the web site wants to have. Most cookies are harmless ways for merchants to know which web pages you've viewed, your preferences, and your credentials. If you're using IE, set your advanced privacy setting to "Prompt" for first-party and third-party cookies. This enables you to decide if you want cookies from a site. For your convenience, you can select the Sites button and use the Per Site Privacy Action option to automatically accept or reject cookies from specific sites.

Many web browsers enable you to store passwords. For maximum security, however, it's better not to use that feature. If you do use that feature, at the Privacy category, go to the subcategory Passwords, and then set a master password to encrypt the data on your computer. This is especially important if you use Mozilla Firefox to manage your passwords. Next, go to the Advanced JavaScript Settings and disable all the options displayed in the dialog box.

Mozilla Firefox

Mozilla Firefox has many of the same features as IE, except the ActiveX and the Security Zone model. To edit the features of Mozilla Firefox, select Tools, and then select Options. Under the General category, you can set Mozilla Firefox as your default web browser. Under the Privacy category, select the Passwords subcategory to manage stored passwords. Then, choose a master password.

Apple Computer's Safari

The Safari contains many of the same features and weaknesses of Mozilla Firefox. To change the Safari settings, click Safari, and then click Preferences. Under the Safari menu, you can choose to block pop-ups. Blocking pop-ups makes your computer more secure, but it may cause you to lose functionality at many sites that use pop-ups to give relevant information.

Other Security Measures

Use an antivirus program. Set it to check your e-mail before you open it. Enable the program to get updated definitions automatically. This lets you keep the antivirus program as current as possible. Also, set your other software programs to receive updates automatically, if possible. When a vendor learns of a vulnerability in its software, it creates patches to make the program more secure.

In many cases, a computer will have multiple web browsers. Even if you only use one, it's important that you configure every web browser for maximum security.

Chapter Quiz

1. Home and business computers are popular targets for computer hackers because they want the information that's stored in them.

   A. True   **B.** False

2. *Spyware* programs are small applications that can get installed on your computer without your knowledge.

   A. True   **B.** False

3. Often, the default settings for web browsers aren't set securely and allow hackers to easily gain control over a computer.

   A. True   **B.** False

4. Many spyware programs can read, write, and delete files, and, in some instances, even reformat your hard drive.

   A. True   **B.** False

5. Because home and small office computers often use high-speed Internet connections that are always turned on, intruders can quickly find and attack them.

   A. True   **B.** False

Chapter 19
TERRORISM

We remain a nation at war. I wish I could report, you know, a different sentence to you. But my job as the President of the United States is to keep the American people fully informed of the world in which we live. In recent months, I've spoken extensively about our strategy for victory in Iraq. Today, I'm going to give you an update on the progress that we're making in the broader war on terror: The actions of our global coalition to break up terrorist networks across the world, plots we've disrupted that have saved American lives, and how the rise of freedom is leading millions to reject the dark ideology of the terrorists—and laying the foundation of peace for generations to come.

President George W. Bush, February 9, 2006

Terrorism is the use or threat of violence to create fear and alarm for political or religious purposes. Terrorists murder and kidnap people, set off bombs, hijack airplanes, set fires, and commit other serious crimes. Despite the huge war machine of the United States, we will never have the firepower to rid the world of terrorism. Terrorists hide in the shadows and lurk in the alleys of the world, and with countless miles of places to hide in, it is impossible to stop all terrorists. However, security professionals and their families, as well as their clients and their clients' families, should have a basic understanding of safety precautions and of the current national antiterrorism initiatives.

Disaster Planning

Terrorism doesn't mean you have to change your life—it just means you need to be prepared. Whether a disaster is natural or human made, being prepared is important. Meet with your family and discuss why you need to be prepared for a disaster and then work together to prepare a family disaster plan.

Discuss the types of hazards that could affect your family. Determine escape routes from your home and places to meet. Pick places for your family to meet outside your home in case of a sudden emergency, such as a fire, or outside your neighborhood if you can't return home.

Have an out-of-state friend or relative as the family contact, so all your family members have a single point of contact. Family members need to call this family contact to let them know where they are in case you cannot be together.

Make a plan now for what to do with your pets if you need to evacuate.

Post emergency telephone numbers by your phones and in your wallet or purse, and make sure your children know how and when to call 911.

Stock nonperishable emergency supplies and a disaster supply kit.

Basic Stock Items

The six basics you should stock for your home are:

1. Water
2. Food
3. First-aid supplies
4. Clothing and bedding
5. Tools and emergency supplies
6. Special items

Keep all your emergency supplies in an easy-to carry container, such as a covered trash container, backpack, or duffle bag.

What Is Suspicious Activity?

Residents may observe a variety of actions, statements, associations, or timing or patterns of activity that create suspicions of illegal conduct. No one has a better perspective about what defines "normal" than the people who live there. Law enforcement has long relied on the common-sense perceptions of citizens who notice something or someone that appears suspicious or out of place.

Who Should I Call to Make a Report?

Call your local police or sheriff or the nearest State Patrol Post. Tell the operator you want to make a suspicious activity report. Ask the operator to alert Homeland Security. Your local law enforcement agency will contact the Sharing and Analysis Center. Agents from the State Bureau of Investigation or the Federal Bureau of Investigation (FBI) will be assigned to carefully check out your information.

Should I Give Police My Name and Contact Number?

Yes! If you want your report to be taken seriously, you should be willing to give your name and contact information to investigators. Someone from State Homeland Security will want to talk to you personally to understand the full details of your information, and then take appropriate action in a timely manner.

Will My Identity Be Protected?

Yes! Reports to State Homeland Security are considered an important part of America's ongoing investigation into the war on terrorism. Investigators need to know your name and contact numbers to do their job, but the state will make every effort to keep your identity confidential.

Do I Have to Talk to the News Media?

No! No one who makes a report to State Homeland Security is required to speak with the news media. State Homeland Security will not release your name to reporters. The decision to remain anonymous to the public or to speak with the news media is left completely up to you.

How Should I Focus My Attention?

Everyone should be especially mindful of suspicious activity around what Homeland Security calls "critical infrastructure." These sites are places or facilities where damage or destruction could cause an interruption of service or result in serious injury or death.

What Should I Watch For?

Citizens should immediately report people who photograph, videotape, sketch, or seek blueprints for dams, drinking water supplies, and water treatment facilities; major highway intersections, bridges, and tunnels; ports, transportation hubs, airports, and shipping facilities; electric plants and substations, and nuclear facilities and transmission towers; pipelines and tank farms; military installations, law enforcement agencies, and defense contract sites; hospitals and health research facilities; Internet, phone, cable, and communications facilities and towers; and capitol, court, and government buildings. Suspicious activity around historic structures and national landmarks also should be reported.

Is My Awareness Really That Important?

Intelligence agents at the State Information Sharing and Analysis Center have investigated an average of one Homeland Security tip every day for nearly a year.

Homeland Security Presidential Directive

The nation requires a Homeland Security Advisory System to provide a comprehensive and effective means to disseminate information regarding the risk of terrorist acts to federal, state, and local authorities, and to the American people. Such a system would provide warnings in the form of a set of graduated "Threat Conditions," which would increase as the risk of the threat increases. At each Threat Condition, federal departments and agencies would implement a corresponding set of "Protective Measures" to further reduce vulnerability or increase response capability during a period of heightened alert.

This system is intended to create a common vocabulary, context, and structure for an ongoing national discussion about the nature of the threats that confront the homeland, as well as the appropriate measures that should be taken in response. It seeks to inform and facilitate decisions appropriate to different levels of government, and to private citizens at home and at work.

Homeland Security

Since September 11, 2001, President Bush has restructured and reformed the federal government to focus resources on counterterrorism and ensure the security of our homeland.

Homeland Security Advisory System

The Homeland Security Advisory System shall be binding on the executive branch and suggested, although voluntary, to other levels of government and the private sector. There are five Threat Conditions, each identified by a description and corresponding color. From lowest to highest, the levels and colors are:

- Low = Green
- Guarded = Blue
- Elevated = Yellow
- High = Orange
- Severe = Red

The higher the Threat Condition, the greater the risk of a terrorist attack. Risk includes both the probability of an attack occurring and its potential gravity, Threat Conditions shall be assigned by the Attorney General in consultation with the Assistant to the President for Homeland Security. Except in exigent circumstances, the Attorney General shall seek the views of the appropriate Homeland Security Principals or their subordinates, and other parties as appropriate, on the Threat Condition to be assigned.

Threat Conditions may be assigned for the entire nation, or they may be set for a particular geographic area or industrial sector. Assigned Threat Conditions shall be reviewed at regular intervals to determine whether adjustments are warranted.

For facilities, personnel, and operations inside the territorial United States, all federal departments, agencies, and offices other than military facilities shall conform their existing threat advisory systems to this system and henceforth administer their systems consistent with the determination of the Attorney General with regard to the Threat Condition in effect.

The assignment of a Threat Condition shall prompt the implementation of an appropriate set of Protective Measures. Protective Measures are the specific steps an organization shall take to reduce its vulnerability or increase its ability to respond during a period of heightened alert. The authority to craft and implement Protective Measures rests with the federal departments and agencies. It is recognized that departments and agencies may have several preplanned sets of responses to a particular Threat Condition to facilitate a rapid, appropriate, and tailored response. Department and agency heads are responsible for developing their own Protective Measures and other antiterrorism or self-protection and continuity plans, and resourcing, rehearsing, documenting, and maintaining these plans. Likewise, they retain the authority to respond, as necessary, to risks, threats, incidents, or events at facilities within the specific jurisdiction of their department or agency, and, as authorized by law, to direct agencies and industries to implement their own Protective Measures. They shall continue to be responsible for taking all appropriate proactive steps to reduce the vulnerability of their personnel and facilities to terrorist attack. Federal department and agency heads shall submit an annual written report to the President, through the Assistant to the President for Homeland Security, describing the steps they have taken to develop and implement appropriate Protective Measures for each Threat Condition. Governors, mayors, and the leaders of other organizations are encouraged to conduct a similar review of their organizations' Protective Measures.

The decision whether to publicly announce Threat Conditions shall be made on a case-by-case basis by the Attorney General in consultation with the Assistant to the President for Homeland Security. Every effort shall be made to share as much information regarding the threat as possible, consistent with the safety of the nation. The Attorney General shall ensure, consistent with the safety of the nation, that state and local government officials and law enforcement authorities are provided the most relevant and timely information. The Attorney General shall be responsible for identifying any other information developed in the threat assessment process that would be useful to state and local officials and others, and conveying it to them as permitted consistent with the constraints of classification. The Attorney General shall establish a process and a system for conveying relevant information to federal, state, and local government officials, law enforcement authorities, and the private sector expeditiously.

The Director of Central Intelligence and the Attorney General shall ensure that a continuous and timely flow of integrated threat assessments and reports is provided to the President, the Vice President, Assistant to the President and Chief of Staff, the Assistant to the President for Homeland Security, and the Assistant to the President for National Security Affairs. Whenever possible and practicable, these integrated threat assessments and reports shall be reviewed and commented upon by the wider interagency community.

A decision on which Threat Condition to assign shall integrate a variety of considerations. This integration will rely on qualitative assessment, not quantitative calculation. Higher Threat Conditions indicate greater risk of a terrorist act, with risk including both probability and gravity. Despite best efforts, there can be no guarantee that, at any given Threat Condition, a terrorist attack will not occur. An initial and important factor is the quality of the threat information itself. The evaluation of this threat information shall include, but not be limited to, the following factors:

1. To what degree is the threat information credible?

2. To what degree is the threat information corroborated?

3. To what degree is the threat specific and/or imminent?

4. How grave are the potential consequences of the threat?

Threat Conditions and Associated Protective Measures

The world has changed since September 11, 2001. We remain a nation at risk to terrorist attacks and will remain at risk for the foreseeable future. At all Threat Conditions, we must remain vigilant, prepared, and ready to deter terrorist attacks. The following Threat Conditions each represent an increasing risk of terrorist attacks. Beneath each Threat Condition are some suggested Protective Measures, recognizing that the heads of federal departments and agencies are responsible for developing and implementing appropriate agency-specific Protective Measures:

1. *Low condition (green).* This condition is declared when there is a low risk of terrorist attacks. Federal departments and agencies should consider the following general measures, in addition to the agency-specific Protective Measures they develop and implement:

   – Refining and exercising as appropriate preplanned Protective Measures; ensuring personnel receive proper training on the Homeland Security Advisory System and specific preplanned department or agency Protective Measures; and institutionalizing a process to assure that all facilities and regulated sectors are regularly assessed for vulnerabilities to terrorist attacks, and all reasonable measures are taken to mitigate these vulnerabilities.

2. *Guarded condition (blue).* This condition is declared when there is a general risk of terrorist attacks. In addition to the Protective Measures taken in the previous Threat Condition, federal departments and agencies should consider the following general measures in addition to the agency-specific Protective Measures that they will develop and implement: checking communications with designated emergency response or command locations; reviewing and updating emergency response procedures; and providing the public with any information that would strengthen its ability to act appropriately.

3. *Elevated condition (yellow).* An Elevated Condition is declared when there is a significant risk of terrorist attacks. In addition to the Protective Measures taken in the previous Threat Conditions, federal departments and agencies should consider the following general measures, in addition to the Protective Measures that they will develop and implement: increasing surveillance of critical locations; coordinating emergency plans as appropriate with nearby jurisdictions; assessing whether the precise characteristics of the threat require the further refinement of preplanned Protective Measures; and implementing, as appropriate, contingency and emergency response plans.

4. *High condition (orange).* A High Condition is declared when there is a high risk of terrorist attacks. In addition to the Protective Measures taken in the previous Threat Conditions, federal departments and agencies should consider the following general measures in addition to the agency-specific Protective Measures that they will develop and implement:

   – Coordinating necessary security efforts with federal, state, and local law enforcement agencies or any National Guard or other appropriate armed forces organizations; taking additional precautions at public events, and possibly considering alternative venues or even cancellation; preparing to execute contingency procedures, such as moving to an alternate site or dispersing their workforce; and restricting threatened facility access to essential personnel only.

5. *Severe condition (red).* A Severe Condition reflects a severe risk of terrorist attacks. Under most circumstances, the Protective Measures for a Severe Condition are not intended to be sustained for substantial periods of time. In addition to the Protective Measures in the previous Threat Conditions, federal departments and agencies also should consider the following gen-

eral measures, in addition to the agency-specific Protective Measures that they will develop and implement:

– Increasing or redirecting personnel to address critical emergency needs; assigning emergency response personnel and prepositioning and mobilizing specially trained teams or resources; monitoring, redirecting, or constraining transportation systems; and closing public and government facilities.

Comment and Review Periods

The Attorney General, in consultation and coordination with the Assistant to the President for Homeland Security, shall, for 45 days from the date of this directive, seek the views of government officials at all levels and of public interest groups and the private sector on the proposed Homeland Security Advisory System.

One hundred thirty-five days from the date of this directive, the Attorney General, after consultation and coordination with the Assistant to the President for Homeland Security, and having considered the views received during the comment period, shall recommend to the President in writing proposed refinements to the Homeland Security Advisory System.

Chapter Quiz

1. Terrorism is the use or threat of violence to create fear and alarm for political or religious purposes.

    A. True   **B.** False

2. If you notice suspicious reports, you should call your local police or sheriff or the nearest State Patrol Post.

    A. True   **B.** False

3. Since September 11, 2001, President Bush has restructured and reformed the federal government to focus resources on counterterrorism and ensure the security of our homeland.

    A. True   **B.** False

Appendix A
ANSWERS TO CHAPTER QUIZZES

Chapter 1 Answers

1. B
2. A
3. A
4. A
5. A
6. A
7. A
8. A
9. A
10. A
11. A
12. A
13. B
14. B
15. A
16. B

Chapter 2 Answers

1. C
2. B
3. A
4. A
5. A
6. A
7. A
8. A
9. A
10. The Wackenhut Corporation
11. A
12. A
13. A
14. A
15. A

Chapter 3 Answers

1. A
2. A
3. A
4. A
5. A
6. A
7. B
8. A
9. Kalamin
10. A
11. B
12. A
13. A
14. A
15. A
16. A
17. A
18. A
19. A
20. A

Chapter 4 Answers

1. A
2. B
3. A
4. B
5. Double-hung
6. Jalousie
7. Casement
8. A
9. A
10. A
11. A

12.	A
13.	A
14.	Lock a double-hung window into place.
15.	B

Chapter 5 Answers

1.	A
2.	A
3.	B
4.	A
5.	A
6.	A
7.	A
8.	B
9.	A
10.	A
11.	A
12.	A
13.	A
14.	B
15.	A

Chapter 6 Answers

1.	A
2.	A
3.	A
4.	B
5.	A
6.	A
7.	B
8.	A
9.	A
10.	A
11.	A

12.	B
13.	A
14.	A
15.	A
16.	A
17.	A
18.	A
19.	A
20.	A

Chapter 7 Answers

1.	The locks often cost from four to ten times more than typical high-security mechanical locks. Many people think electromagnetic locks are less attractive than mechanical locks.
2.	B
3.	A
4.	A
5.	A
6.	A
7.	A
8.	A
9.	A
10.	A

Chapter 8 Answers

1.	A
2.	A
3.	B
4.	A
5.	A
6.	A
7.	A
8.	A

9. A

10. A

Chapter 9 Answers

1. A
2. A
3. A
4. B
5. A
6. A
7. A
8. A
9. A
10. A

Chapter 10 Answers

1. B
2. A
3. A
4. B
5. A
6. B
7. A
8. A
9. A
10. A
11. A
12. A
13. B
14. A
15. A

Chapter 11 Answers

1. A

2. A
3. A
4. A
5. A

Chapter 12 Answers

1. B
2. B
3. B
4. B
5. B

Chapter 13 Answers

1. A
2. A
3. A
4. A
5. A
6. A
7. B
8. A
9. A
10. A

Chapter 14 Answers

1. A
2. B
3. B
4. B
5. A
6. A
7. A
8. A

9. A

10. B

11. A

12. A

13. A

14. A

15. A

16. B

17. A

18. A

19. A

20. A

21. A

22. A

Chapter 15 Answers

1. A

2. A

3. A

4. B

5. A

6. A

7. A

8. A

9. A

10. A

Chapter 16 Answers

1. A

2. A

3. A

4. A

5. A

Chapter 17 Answers

1. A

2. A

3. A

4. A

5. B

Chapter 18 Answers

1. A

2. A

3. A

4. A

5. A

Chapter 19 Answers

1. A

2. A

3. A

ASSOCIATED LOCKSMITHS OF AMERICA GENERAL LOCKSMITH CERTIFICATION EXAM

Note: The following exam includes sample questions from the General Locksmith Certification Exam required for the following designations: Registered Locksmith (RL), Certified Registered Locksmith (CRL), Certified Professional Locksmith (CPL), Certified Master Locksmith (CML), Certified Professional Safe Tech (CPST), and Certified Master Safe Tech (CMST). In addition to the General Locksmith Certification Exam, elective exams are required.

1. Key code A*51 in a code book marked with a note of "– 450=DB" is:

   A. A direct digit

   B. A derivative or conversion

   C. The same code listed as DB-A451

   D. None of the above

2. How are codes listed in code books?

   A. From bow to tip

   B. From tip to bow

   C. It depends on the code book.

   D. It depends on the country of origin.

3. Although automotive keys are often punched out on "code clippers," many other keys like Schlage, Best, and Arrow cannot be punched accurately.

   A. True   B. False

4. Disc tumbler locks are master keyed by:

   A. Inserting more than one disc in each chamber.

   B. Cutting back the tip of the change key.

   C. Filing the tumblers, so different keys will work.

   D. Using stepped tumblers.

   E. None of the above

5. The ALOA Glossary preferred term for "wafer tumbler" is:

   A. Disc tumbler

   B. Shim tumbler

   C. Series tumbler

   D. Plate tumbler

6. The key section drawing underneath the key blank illustration in an after-market key blank catalog is normally the key blade viewed from the tip.

   A. True   **B.** False

7. To duplicate a key for a lever-tumbler safe deposit lock, you would normally use a rotary file cutter.

   A. True   **B.** False

8. Wrenching a cylinder from a mortise deadlock will probably destroy:

   A. The cylinder

   B. The cylinder and damage the lock

   C. Only the cylinder set screw

   D. None of the above

9. Which lockset function from the following listing would be described as "outside knob always locked, entry by key only, inside knob always free"?

   A. Entry

   B. Storeroom

   C. Classroom

   D. Vestibule

10. The BHMA standard number for a finish described as "Oil Rubbed Bronze" is:

    A. 605

    B. 609

    C. 612

    D. 613

11. Which lockset must be removed from the door to release the cylinder?

    A. Arrow Mil

    B. Schlage A53

    C. Kwikset 400

    D. Best

    E. None of the above

12. Using the Standard Key Coding System, which of the following key symbols is from a simple two-level master-key system?

   A. Al A

   B. AAl

   C. 23 AA

   D. XAA1

13. Letter box locks utilizing a flat steel key are:

   A. Pin tumbler

   B. Lever tumbler

   C. Disc tumbler

   D. Warded

14. The technique of combining several alarm signals over a single transmission medium is known as:

   A. Direct wire

   B. Line security

   C. Multiplexing

   D. Digital dialer

15. UL standards require alarm circuit wires that are run in the vicinity of power lines:

   A. Cross at 90 degrees and be separated by at least 2 inches.

   B. Run parallel and be separated by at least 2 inches.

   C. Be separated by an insulator.

   D. Be rated as a Class 2 circuit.

   E. None of the above

16. In the Ford products using the ten-cut system, how many cuts on the key are common to both the door and ignition cylinders?

   A. 2

   B. 4

   C. 6

   D. 10

17. Chrysler vehicles utilizing air bags will have double-sided keys.

   A. True   **B.** False

18. It is possible to fit a key by reading the disc tumblers through the keyway of most imported car locks without picking the cylinder.

   A. True   **B.** False

19. Pebra dimple key cylinders may be found in the ignition of which of the following automobiles?

   A. Rolls Royce

   B. DeLorean

   C. Peugeot

   D. BMW

20. All door closers can be rebuilt in the field.

   A. True   **B.** False

21. Which of the following is considered by NFPA 101 as a "place of assembly"?

   A. A retail store

   B. A classroom

   C. A theater

   D. A hotel suite

22. Which of these manufacturers has not furnished a high-security interchangeable core cylinder?

   A. Abloy

   B. KABA

   C. Emhart

   D. Dom

23. The best choice of a lock for use in a high-humidity area would be a Schlage:

   A. A series

   B. B series

   C. C series

   D. D series

24. Which lock manufacturer does not provide a UL-listed cylinder?

    A. Fort

    B. National

    C. Hudson

    D. Chicago

25. Burglar-alarm circuits are classified under which article of the National Electric Code?

    A. Article 800

    B. Article 725

    C. Article 760

    D. Article 810

26. Which of the following manufacturers does not provide an exit alarm that locks the door?

    A. DADCO

    B. Alarm Lock

    C. Best

    D. Detex

27. The shipping combination of a Simplex/Unican lock is:

    A. 2 and 4 together, and then 3

    B. 2 and 4 together, and then 1

    C. 1, 2, 3

    D. 1, and then 3 and 5 together

28. It is possible to set the combination on a Sentry Safe model 6310 to a set of numbers chosen by the customer.

    A. True   B. False

29. The most common reason for a lockout on a safe or vault equipped with a time lock is:

    A. Binding pressure on the master lever in the time lock.

    B. All of the clocks have stopped.

    C. The time lock is overwound.

    D. The timer is holding the combination lock bolt.

30. A door that is factory prepared with a "160 prep" would use a:

    A. Standard duty cylindrical lockset

    B. Unit lockset, classroom function

    C. Sargent 8-Line

    D. Pocket door lock

31. Using the Standard Key Coding System, the keyset "A2A49" would correspond to:

    A. A cross-keyed cylinder.

    B. The forty-ninth change key under the second GMK with no master key to operate.

    C. A change key in a system with at least three levels of keying.

    D. Nothing in the standard key coding system.

32. The term "live load" refers to the maximum load:

    A. That can be placed in a 2' × 2' space.

    B. That can be placed on a floor other than the building materials.

    C. On a floor as it relates to the occupancy level.

    D. At the point equidistant from any three support columns.

33. Manipulation is accomplished by dial readings taken at the:

    A. Contact points

    B. Changing index

    C. Drill point

    D. Fence location

34. What is the most common reason for a bank vault lockout?

    A. Weak time lock spring

    B. Dirty carry bar

    C. Incorrect winding

    D. All of the above

35. Detention locks may use lever-tumbler or pin-tumbler mechanisms.

    A. True   **B.** False

Appendix C

CERTIFIED PROTECTION PROFESSIONAL EXAM

Locksmith and Security Professionals' Exam Study Guide

1. When a security officer has been commissioned with special police or peace officer powers, this is usually unlikely to hold up in court.

   A. True   **B.** False

2. When a security officer has been commissioned with special police or peace officer powers, this is usually limited to the grounds and buildings of the officer's employer.

   A. True   **B.** False

3. When a security officer has been commissioned with special police or peace officer powers, it usually applies statewide, but only for a limited amount of time.

   A. True   **B.** False

4. In a large security firm, the role of an ombudsman is to coordinate financial policy.

   A. True   **B.** False

5. In a large security firm, the role of an ombudsman is to handle employee grievances and ethical problems.

   A. True   **B.** False

6. The three types of evidence are photographed, taped, and spoken.

   A. True   **B.** False

7. The three types of evidence are sworn, eyewitness, and hearsay.

   A. True   **B.** False

8. The three types of evidence are narrative, testimonial, and circumstantial.

   A. True   **B.** False

9. The three types of evidence are testimonial, documentary, and real.

   A. True   **B.** False

10. Ecstasy is the most widely used illegal drug.

    A. True   **B.** False

11. Marijuana is the most widely used illegal drug.

    A. True   **B.** False

12. Methamphetamine is the most widely used illegal drug.

    A. True   **B.** False

13. Cocaine is the most widely used illegal drug.

    A. True   **B.** False

14. Regardless of whether a business is product- or service-oriented, the initial requirement in creating a loss-prevention program is to identify the business purpose.

    A. True   **B.** False

15. Regardless of whether a business is product- or service-oriented, the initial requirement in creating a loss-prevention program is to define protection activities.

    A. True   **B.** False

16. Regardless of whether a business is product- or service-oriented, the initial requirement in creating a loss-prevention program is to identify specific vulnerabilities.

    A. True   **B.** False

17. Regardless of whether a business is product- or service-oriented, the initial requirement in creating a loss-prevention program is to define protection objectives.

    A. True   **B.** False

18. At most government agencies, if an Executive Director suddenly becomes absent, the Program Director automatically steps into the role of Acting Executive Director.

    A. True   **B.** False

19. At most government agencies, if an Executive Director suddenly becomes absent, the Deputy Director automatically steps into the role of Acting Executive Director.

    A. True   **B.** False

20. At most government agencies, if an Executive Director suddenly becomes absent, the Ombudsman automatically steps into the role of Acting Executive Director.

    A. True   **B.** False

21. At most government agencies, if an Executive Director suddenly becomes absent, the Chief Financial Officer automatically steps into the role of Acting Executive Director.

   A. True   **B.** False

22. In criminal investigations, physical specimens are known as artifacts.

   A. True   **B.** False

23. In criminal investigations, physical specimens are known as testimonials.

   A. True   **B.** False

24. In criminal investigations, physical specimens are known as examplars.

   A. True   **B.** False

25. In addition to solid, ethical work, an officer's best defense against a lawsuit is a testimonial from a trusted character witness.

   A. True   **B.** False

26. In addition to solid, ethical work, an officer's best defense against a lawsuit is clear, detailed records and reports.

   A. True   **B.** False

27. In most burglaries, the most likely suspects are employees.

   A. True   **B.** False

28. In most burglaries, the most likely suspects are family members.

   A. True   **B.** False

29. In most burglaries, the most likely suspects are friends.

   A. True   **B.** False

30. In most burglaries, the most likely suspects are one-time visitors.

   A. True   **B.** False

31. After a crime is committed, a security officer's role is usually apprehension and detention.

   A. True   **B.** False

32. After a crime is committed, a security officer's role is usually observing and reporting.

A. True   **B.** False

33. Most commercial break-ins are done through adjacent buildings.

A. True   **B.** False

34. Most commercial break-ins are done through doors and windows.

A. True   **B.** False

35. In most statement reports, the final component is the attestation.

A. True   **B.** False

36. In most statements reports, the final component is the notary seal.

A. True   **B.** False

37. The typical holding force of an electromagnet lock is about 4,000 lbs.

A. True   **B.** False

38. The typical holding force of an electromagnet lock is about 750 pounds.

A. True   **B.** False

39. The typical holding force of an electromagnet lock is about 1,500 pounds.

A. True   **B.** False

40. If a private security officer comes across a crime scene, they should collect and bag physical evidence.

A. True   **B.** False

41. If a private security officer comes across a crime scene, they should begin canvassing for witnesses.

A. True   **B.** False

42. If a private security officer comes across a crime scene, they should call the police.

A. True   **B.** False

43. A change key is made to fit only one lock.

    A. True   **B.** False

44. Often, when a criminal act causes a loss for a business, judges require only the full replacement value of the tangible property to be associated with prosecution.

    A. True   **B.** False

45. Often when a criminal act causes a loss for a business, judges require only the retail value of the tangible property to be associated with prosecution.

    A. True   **B.** False

46. Often, when a criminal act causes a loss for a business, judges require only the cost value of the tangible property to be associated with prosecution.

    A. True   **B.** False

47. Security consoles in a large installation shouldn't be conveniently located.

    A. True   **B.** False

48. Logic bombs are a type of computer virus frequently used for embezzlement.

    A. True   **B.** False

49. Trojan horses are a type of computer virus frequently used for embezzlement.

    A. True   **B.** False

50. The "Standard Checklist Criteria for Business Recovery," published by the Federal Emergency Management Agency (FEMA), recommends that a company's business recovery plan be updated every year.

    A. True   **B.** False

51. The financial planning process at a security organization typically doesn't include budgeting.

    A. True   **B.** False

52. If a suspect perspired a lot during questioning, they will be described by a forensic technologist as a "secretor."

    A. True   **B.** False

53. In a management scheme that follows Hertzberg's motivational model, a sense of responsibility would serve as the strongest motivating factor for an employee.

   A. True   **B.** False

54. Negative reinforcement occurs when an employee's behavior is accompanied by the removal of an unfavorable consequence.

   A. True   **B.** False

55. The main disadvantage of most wireless alarm systems is they are harder to install than hard-wired systems.

   A. True   **B.** False

56. Before a crime is committed, a security officer's role is prevention.

   A. True   **B.** False

57. Heroin is the most widely used synthetic opiate.

   A. True   **B.** False

58. Management implications inherent in MacGregor's Theory Y of management don't include decentralization.

   A. True   **B.** False

59. When writing an investigative report, a security officer shouldn't compose the report in chronological order.

   A. True   **B.** False

60. A night latch has a spring latch bolt and provides extra strength to a door's primary lock.

   A. True   **B.** False

61. When investigating a crime, an officer should return to reinterview a witness who could be described as talkative.

   A. True   **B.** False

62. The most appropriate process for conducting a crime scene over a large area is the spiral method.

   A. True   **B.** False

63. In most jurisdictions, a security officer has the same power to arrest as a private citizen.

    A. True   **B.** False

64. When a hardwired alarm system is experiencing many false alarms, the cause is often a broken wire or a loose connection.

    A. True   **B.** False

65. If the security manager is working in a proprietary system, they are most likely to report to the Chief Executive Officer.

    A. True   **B.** False

66. According to the Hallcrest Report on private security and investigation, employee theft accounts for about 90 percent of a company's losses.

    A. True   **B.** False

67. Within a large security organization, tactical or operational problems are primarily the responsibility of a mid-level manager.

    A. True   **B.** False

68. Many private computer-system breaches can be prevented by installing correct software patches and security upgrades.

    A. True   **B.** False

69. When taken from an automobile as evidence, paint chips can often be used to identify the vehicle's year of manufacturer.

    A. True   **B.** False

70. The most common type of security lighting system is the standby system.

    A. True   **B.** False

71. A security manager who adopts a "custodial" leadership style bases their decisions on leadership.

    A. True   **B.** False

72. When a security officer is helping in the investigation of a crime scene that may contain relevant fingerprints, the officer should dust the item for prints.

    A. True   **B.** False

73. "Avoidance" is a term for a risk response that involves eliminating a threat.

    A. True   **B.** False

74. "Mitigation" is a term for a risk response that involves eliminating a threat.

    A. True   **B.** False

75. An organization is practicing "active risk acceptance" when it develops a plan to minimize probability.

    A. True   **B.** False

76. Most narcotics are derived from opium.

    A. True   **B.** False

77. Most narcotics are derived from hemp.

    A. True   **B.** False

78. The normal risk of doing business that carries opportunities for both gain and loss is called favorable risk.

    A. True   **B.** False

79. The normal risk of doing business that carries opportunities for both gain and loss is called opportunity risk.

    A. True   **B.** False

80. A security manager who adopts an "autocratic" leadership style bases their decisions on formal lines of authority.

    A. True   **B.** False

81. A security manager who adopts an "autocratic" leadership style bases their decisions on teamwork.

    A. True   **B.** False

82. Window location is classified as a "natural" security strategy.

    A. True   **B.** False

83. Most coding systems for master key combinations are based on differences in tumbler depth.

    A. True   **B.** False

84. Most coding systems for master key combinations are based on differences in key blade width.

    A. True   **B.** False

85. The diversion of resources and assets to lower loss exposure is called transfer.

    A. True   **B.** False

86. The diversion of resources and assets to lower loss exposure is called abatement.

    A. True   **B.** False

87. Usually a good place to get a sample of a person's signature is from records at the department of motor vehicles.

    A. True   **B.** False

88. An evidentiary item's chain of custody is usually considered to have been initiated by either the recovering officer or the reporting officer.

    A. True   **B.** False

89. An evidentiary item's chain of custody is usually considered to have been initiated by the witness who first found it.

    A. True   **B.** False

90. An evidentiary item's chain of custody is usually considered to have been initiated by the property or evidence specialist.

    A. True   **B.** False

91. One of the biggest advantages associated with the use of firewalls in a computer network security is the entire network presents only one IP address to the outside world.

    A. True   **B.** False

92. The bottom-up method of budgeting works best when competitive pressures require a quick response.

    A. True   **B.** False

93. The bottom-up method of budgeting works best when first-line management is excluded from the process.

    A. True   **B.** False

94. A dead-latch locking bolt is protected by a device that automatically blocks spring action after the door has been shut, preventing jimmying of the bolt.

    A. True   **B.** False

95. A night-latch locking bolt is protected by a device that automatically blocks spring action after the door has been shut, preventing jimmying of the bolt.

    A. True   **B.** False

96. The most important classifying characteristic of fiber evidence is its density.

    A. True   **B.** False

97. The most important classifying characteristic of fiber evidence is its volume.

    A. True   **B.** False

98. The most important classifying characteristic of fiber evidence is its mass.

    A. True   **B.** False

99. "High-security" combination padlocks operate on a 4-3-2-1 rotation.

    A. True   **B.** False

100. "High-security" combination padlocks operate on a 1-2-3-4 rotation.

    A. True   **B.** False

101. "High-security" combination padlocks operate on a 4-4-3-2 rotation.

    A. True   **B.** False

Appendix D

INTERNATIONAL FOUNDATION FOR PROTECTION OFFICERS CERTIFIED PROTECTION OFFICER INTERIM EXAMINATION

Locksmith and Security Professionals' Exam Study Guide

Exam 1

Unit One

History of Law and Security

1. There is no evidence of any written law until 325 A.D.

    A. True   **B.** False

2. The term "curia regis" means King's Court.

    A. True   **B.** False

3. The Middle Ages outshone any other era in the number of revolutionary and significant advances made in the development of legal concepts that have survived to modern day.

    A. True   **B.** False

4. The military services of the Allied Nations in WWII played only a minor role in the development of the private security industry.

    A. True   **B.** False

5. Training of private security personnel, while important, is not the major factor of concern when analyzing the challenges and future of the private security industry.

    A. True   **B.** False

6. The earliest law was:

    A. Written

    B. Passed by word of mouth

    C. Known only to a select few

    D. Both criminal and civil

7. The Norman conquest of England produced very significant developments that impacted on the legal system, including:

    A. The introduction of feudalism

    B. The centralization of government

    C. The reorganization of the church

    D. All the above

8. The 1600s saw:

   A. The abolition of the Star Chamber

   B. The passing of the Habeas Corpus Act

   C. The beginning of the private police to protect merchants' property

   D. All the above

9. Police departments were established in the U.S. cities of New York, Chicago, Philadelphia, and Detroit during:

   A. The 1700s

   B. The 1800s

   C. The 1900s

   D. None of the above

10. The real source and stimulus of the modern private security industry was:

   A. World War I

   B. The industrial revolution

   C. The postwar crime wave and white collar crime

   D. World War II

Field Notes and Report Writing

11. An officer may be requested to show their notebook, at some point, to:

   A. Their previous supervisor

   B. The courts

   C. A competitor

   D. All of the above

   E. None of the above

12. Which of the following should an officer have readily available when writing reports?

   A. A coffee

   B. Extra forms

   C. A dictionary

   D. All of the above

   E. None of the above

13. Security reports could be viewed by:

    A. A judge

    B. A defense lawyer

    C. A security manager

    D. All of the above

    E. None of the above

14. After reading your daily reports, which individual or group would most likely benefit from the contents?

    A. The public

    B. The private justice system

    C. Your fellow officers

    D. All of the above

    E. None of the above

15. When writing an accurate report, it is imperative to refer to:

    A. A phone book

    B. Your notes

    C. Previous reports

    D. All of the above

    E. None of the above

16. Proper notes are the first step in forming a permanent record of events as they occurred.

    A. True   **B.** False

17. Notes are not an essential part of proper report preparation.

    A. True   **B.** False

18. The officer's ability to patrol is judged solely on the basis of their notebook.

    A. True   **B.** False

19. Guidelines for the use of the notebook and report writing are basically the same.

    A. True   **B.** False

20. An officer must be able to go back to their notes and be able to determine the best possible suspect.

    A. True   **B.** False

Observation and Memory

21. The smaller the object, the farther away the observer will be able to recognize it.

    A. True   **B.** False

22. Under normal conditions of visibility, a person with distinctive features can be recognized by friends and relatives at:

    A. 75 yards

    B. 50 yards

    C. 100 yards

    D. 125 yards

23. Which substance may kill your sense of smell temporarily?

    A. Gunpowder

    B. Wood smoke

    C. Ether

    D. Electric smoke

24. When we use our senses effectively, we are thinking and being aware.

    A. True   **B.** False

25. Each person's ability to recall information from memory is the same, regardless of the amount of practice.

    A. True   **B.** False

26. When patrolling, you should stop occasionally just to listen.

    A. True   **B.** False

27. To help improve your sight, you should:

    A. Be aware of what you look at

    B. Insure your vision is in peak condition

    C. Insure you understand the factors that affect your vision

    D. All of the above

28. The position of the observer in relation to the subject can alter the observer's perception of the subject.

    A. True   B. False

29. Your sense of touch assists your job in which of the following ways? (Mark the incorrect answer.)

    A. Feeling walls or glass for unseen heat

    B. Checking the consistency of a substance

    C. Checking mufflers for warmth

    D. Checking the pulse of an accident victim

30. In administering duties, which sense would almost never be used?

    A. Touch

    B. Sight

    C. Taste

    D. Smell

    E. All of the above

Unit Two

Patrol Techniques

31. Which of the following can enhance the patrolman's skill and ability? (Select the best answer.)

    A. Proper training

    B. Preparation for patrol

    C. Professional work habits

    D. All of the above

32. Security officers should act and look professional only while at work.

    A. True   B. False

33. Detecting criminals is the major Security Officer responsibility.

    A. True   **B.** False

34. Under no circumstance should an officer leave the post until properly relieved.

    A. True   **B.** False

35. It is not essential that the Security Officer document all observations.

    A. True   **B.** False

36. Based on organizational needs, there are several major purposes of patrol. (Choose the best answer.)

    A. Respond to emergencies

    B. Prevention and deterrence of crime

    C. Detection of criminal activity

    D. All the above

37. The function of security is to prevent and control loss.

    A. True   **B.** False

38. The WAECUP Theory includes which of the following?

    A. Waste

    B. Error

    C. Accident

    D. Crime

    E. All of the above

39. The advantage(s) of mobile patrols is/are: (Choose the best answer.)

    A. Coverage of more extensive areas

    B. More comfortable for the guards involved

    C. Able to better observe surroundings

    D. Acts as a better deterrent to the general public

    E. All of the above

40. The first principle of patrol is that it should always be done in a random fashion.

    A. True   **B.** False

Safety and the Protection Officer

41. Which of the following safety/security conditions should be of concern to the Protection Officer on patrol? (Mark the item that is out of place.)

    A. Boxes blocking an exit

    B. A flaw in computer interfacing

    C. An obstructed CCTV camera lens

    D. An overheating electrical motor

42. Unsafe conditions, such as poor housekeeping, can be a major contributor to an accident.

    A. True   **B.** False

43. Accident prevention measures taken by the Protection Officer while on patrol relate to: (Mark the best answer.)

    A. Taking an active role in Safety Committee meetings

    B. Monitoring employee behavior

    C. Noting and reporting safety hazards

    D. Providing meaningful reports that can be interpreted by top management

44. Employee training has a positive effect in terms of developing on-the-job safety practices.

    A. True   **B.** False

45. Providing direct employee safety training is the responsibility of: (Mark the best answer.)

    A. Manager

    B. Supervisor

    C. Safety Committee

    D. Protection Officer

46. Safety meetings are exclusively for the Safety Committee.

    A. True   **B.** False

47. Safety contests with awards are a good way to improve safety practices by employees because:

   A. This practice resembles a lottery.

   B. This practice gets employees involved with managers.

   C. This practice increases employee safety awareness and motivation.

   D. This practice is proven effective in strengthening union-management relations.

48. Safety posters have been proven as the best method of conveying employee safety-awareness practices.

   A. True   **B.** False

49. The safety committee's investigation of a fatal accident is conducted with a view to: (Mark the best answer.)

   A. Determining if criminal charges are involved

   B. Establishing if there has been any "horse-play" on the part of employees

   C. Determining the cause to prevent future accidents

   D. Finding out who is to blame for the mishap

50. Drug abuse in the workplace is the number one cause of accidents.

   A. True   **B.** False

Traffic Control

51. A corner position provides a better view of traffic than a center-of-the-intersection position.

   A. True   **B.** False

52. When traffic is congested and motorists desire frequent turns that result in a slowdown of traffic flow, they should be:

   A. Stopped

   B. Obliged

   C. Forced to go straight through

   D. Pulled over to give the intersection a chance to clear

53. Proper officer protection against the elements is an important factor in maintaining efficient traffic control.

   A. True   **B.** False

54. Prompt compliance to hand signals is dependent on the officer's ability to:

    A. Use uniform, clearly defined hand signals

    B. Quickly assess traffic flow needs

    C. Quickly assess congestion problems

    D. All of the above

55. When an emergency vehicle is approaching, you stop:

    A. All pedestrian traffic

    B. Only vehicles on the street on which the emergency vehicle is approaching

    C. The emergency vehicle

    D. All vehicular and pedestrian traffic

56. When directing traffic, priority of movement is determined by the amount of traffic flow in each direction.

    A. True   **B.** False

57. When attempting to attract a motorist's attention with a whistle, give:

    A. One long blast

    B. Two long blasts

    C. Two short blasts

    D. One short blast

58. The property owner of a private parking lot is responsible for controlling traffic and patrolling the area.

    A. True   **B.** False

59. Traffic control at construction sites is only for the protection of the workers.

    A. True   **B.** False

60. Proper direction of traffic:

    A. Regulates the flow of traffic

    B. Protects pedestrians

    C. Assists emergency vehicles

    D. All of the above

Crowd Control Management/Procedures

61. A crowd may initially exist as a casual or temporary assembly having no cohesiveness.

    A. True   **B.** False

62. It is important for the Protection Officer to quickly determine if a gathering may become uncontrollable.

    A. True   **B.** False

63. Members of a crowd who assemble for a sporting event depend on each other for support and have a unity of purpose.

    A. True   **B.** False

64. Persons joining a crowd tend to accept the ideas of its dominant members without realization or conscious objection.

    A. True   **B.** False

65. An individual can lose self-consciousness and identity in a crowd.

    A. True   **B.** False

66. There is potential for mass discord whenever people gather at:

    A. Athletic events

    B. Parades

    C. Protest rallies

    D. All the above

    E. None of the above

67. In dealing with persons in a crowd situation, a Protection Officer should:

    A. Use good judgment and discretion

    B. Remain impartial and courteous

    C. Refrain from derogatory remarks

    D. All of the above

    E. None of the above

68. Emotional reactions resulting in the formation of unruly crowds are often associated with:

    A. Absence of authority

    B. Religious/racial differences

    C. Economic conditions

    D. All of the above

    E. None of the above

69. A hostile crowd is usually:

    A. Motivated by feelings of hate

    B. Gathered to watch out of interest

    C. Attempting to flee from something it fears

    D. All of the above

    E. None of the above

70. Psychological causes for crowd formation can be:

    A. Feeling of security

    B. Loss of identity

    C. Release of emotions

    D. All of the above

    E. None of the above

Crime Scene Procedures

71. It is important for the Protection Officer to summon enough assistance to:

    A. Properly protect the crime scene

    B. Offer advice on correct procedures

    C. Help gather vital evidence

    D. All of the above

    E. None of the above

72. The reason the Protection Officer must quickly attend all crime scenes is:

    A. To preserve all possible evidence

    B. To show the client they are efficient

    C. To quickly give chase to the culprit

    D. All of the above

73. The moment an officer arrives at a crime scene, they should consider:

    A. Precautions to ensure personal safety

    B. Injured persons

    C. Notes/information

    D. All of the above

    E. None of the above

74. Items that should be recorded in the officer's notebook at a crime scene include:

    A. The date and time of the officer's arrival

    B. Persons present

    C. Date and time of occurrence

    D. All of the above

    E. None of the above

75. A reconstruction of a crime scene could:

    A. Show the crime was not actually committed

    B. Lead to further evidence

    C. Indicate clues to culprit identity

    D. All of the above

    E. None of the above

76. The main purpose of collecting evidence is to: (Mark the best answer.)

    A. Enter it as evidence

    B. Aid in the identification and conviction of the accused

    C. Ensure that curious persons do not contaminate it

    D. All of the above

    E. None of the above

77. Once the boundaries of a crime scene have been established, it is less important to keep people away from the area.

    A. True   B. False

78. Fellow officers and occupants at the crime scene need not be excluded from the scene once the boundaries have been established.

    A. True   B. False

79. Once the area of a crime scene has been established, it is unnecessary to exceed these boundaries in the search for evidence.

    A. True   **B.** False

80. The Protection Officer is usually the first person to come upon a crime scene.

    A. True   **B.** False

Unit Three

Physical Security System

81. The final step in the process of conducting a security vulnerability analysis is the identification of assets.

    A. True   **B.** False

82. Security lighting is designed solely to detect intruders.

    A. True   **B.** False

83. Wired glass is considered burglar- and vandal-resistant.

    A. True   **B.** False

84. Audio sensors are motion sensors.

    A. True   **B.** False

85. Ultrasonic detectors are not recommended for use in an area that may be subjected to air turbulence.

    A. True   **B.** False

86. The optical coded badge can be recognized by:

    A. Its shape

    B. The solid black bar across the back of the card

    C. The holes punched in the card

    D. The use of the bar code

87. Closed-circuit television sequential switchers are highly recommended because:

   A. They stamp the date and time on each video tape.

   B. They allow for pan-and-tilt control of cameras.

   C. They allow one monitor to be used with several cameras.

   D. They allow an operator to zoom in for a close-up.

88. The minimum recommended gauge for chain-link fence fabric is:

   A. 6 gauge

   B. 14 gauge

   C. 9 gauge

   D. 12 gauge

89. The top overhang of a fence should increase the height of the fence by at least:

   A. 9 inches

   B. 12 inches

   C. 24 inches

   D. 16 inches

90. Quartz lamps emit a(n):

   A. Bluish light

   B. Yellow light

   C. White light

   D. Orange light

Unit Four

Bomb Threats

91. Knowledge in use and handling of explosives is readily available through book stores and mail-order firms.

   A. True   **B.** False

92. The Bomb Incident Plan (BIP) is a step-by-step instruction on how to handle an explosive device.

   A. True   **B.** False

93. When searching for hidden explosive devices, open cabinet drawers quickly.

    A. True   **B.** False

94. Time permitting, if an explosive device is found, it should be surrounded with sandbags or mattresses.

    A. True   **B.** False

95. One weapon commonly used by terrorists is hand grenades.

    A. True   **B.** False

96. Most bomb threats are a hoax, but:

    A. 5 to 10 percent are the real thing

    B. 10 to 20 percent are the real thing

    C. 20 to 30 percent are the real thing

    D. 50 percent are the real thing

97. High explosives are usually detonated by a blasting cap; however, explosives can also be detonated by:

    A. A time-safety fuse

    B. Gasoline

    C. Matches

    D. Detonation/priming cord

98. In a nonelectric firing system, the activation source would usually be:

    A. Heat or flame

    B. AC or DC current

    C. Blasting machine

    D. None of the above

99. When handling a bomb threat on the telephone:

    A. Warn the caller of existing BIP

    B. Hang up; most bomb threats are hoaxes

    C. Keep the caller talking to receive as much information as possible

    D. Tell the caller to notify the police

100. If a written threat is received:

   A. Throw it out; most written threats are a hoax

   B. Don't take it lightly; any bomb threat is serious

   C. Save all material, along with envelopes or containers

   D. B and C above

Alarm Systems

101. Different alarm systems are incompatible and should never become part of an integrated system.

   A. True   **B.** False

102. Laser models of photoelectric cell alarm systems are a very stable alarm sensor.

   A. True   **B.** False

103. When a photocell alarm system will not "arm," this is most likely caused by:

   A. Improper installation

   B. A blockage of the beam

   C. Cut wires

   D. None of the above

104. An ultrasonic alarm system:

   A. Detects sound above the range of the human ear

   B. Detects sounds below the range of the human ear

   C. Detects only unusual sounds

   D. All of the above

105. Ultrasonic alarm systems cannot detect air turbulence.

   A. True   **B.** False

106. An ultrasonic alarm system, utilizing radio frequencies, can be used in penetrating walls and ceilings to provide better protection.

   A. True   **B.** False

107. A microwave alarm system can "see" inside a heater to determine if the fan is in motion.

   A. True   **B.** False

108. Infrared sensors:

    A. Are state of the art

    B. Present few false alarms

    C. Can detect fire

    D. All of the above

109. A UL-approved alarm:

    A. Must be manned at all times

    B. Must meet company standards

    C. Must have direct communication lines

    D. B and C

110. To properly protect double-paneled sliding glass doors, it is best to choose a combination of:

    A. A microwave and infrared

    B. Infrared and window bugs

    C. Window bugs and door contacts

    D. Door contacts and ultrasonic

Fire Prevention

111. Soda-acid extinguishers contain:

    A. A small bottle of acid

    B. All water

    C. Antifreeze supply

    D. Carbon dioxide

112. Materials termed fire-resistant can sometimes burn if temperatures are extremely high.

    A. True   **B.** False

113. There are two common types of halon extinguishers, which are:

    A. Halon 1110

    B. Halon 1304

    C. Halon 1301

    D. Halon 1211

114. Halon systems are frequently used as a fire-eliminating process in computer centers because "postfire" damage is less than what is caused by other methods of fire extinguishing.

   A. True   **B.** False

115. In the event of a fire that you choose to extinguish, your first action should be:

   A. Activate the extinguisher

   B. Direct the extinguisher as per instructions

   C. Select the correct extinguisher

   D. Keep the exit path clear

116. As soon as you feel a fire is under control, deactivate the extinguisher in case the remainder is needed for a fresh fire.

   A. True   **B.** False

117. The multipurpose range of a dry chemical extinguisher is:

   A. 1M to 2M

   B. 3M to 4.5M

   C. 5M to 7.5M

   D. 7M to 9.5M

118. Foam extinguishers eliminate fire by lowering the temperature.

   A. True   **B.** False

119. Sprinkler system failure is usually attributed to:

   A. Prematurely melting solder on a sprinkler head

   B. An inadequate supply of CO_2

   C. Human error

   D. A foreign substance

120. Flammable liquid fires can be extinguished by air exclusion.

   A. True   **B.** False

Hazardous Materials

121. The majority of chemicals and other substances considered "hazardous materials":

    A. Are controlled by national laws

    B. Must be transported improperly

    C. Are not inherently dangerous in their original state

    D. Are designed to be out of the reach of children

122. Ultimately, all uncontrolled releases can be traced to:

    A. Equipment failure

    B. Human failure

    C. Improperly followed safety procedures

    D. Lack of hazardous material facilities

123. For decades, the common method of response to a hazardous material release was to:

    A. Notify the local fire department or plant fire brigade

    B. Call the local police department

    C. Wash the contaminated area

    D. Get as much citizen involvement as possible

124. The term "Site Security" refers to:

    A. Sealing off the area, pending an investigation of the incident

    B. Keeping onlookers and bystanders out of the contaminated area

    C. Designating simple entry and exit points

    D. Security that prevents spills

125. The highest area of contamination is called:

    A. The hot zone

    B. The contamination reduction zone

    C. The exclusion zone

    D. The critical zone

126. Nonessential personnel may be allowed at the command post.

    A. True   **B.** False

127. The entire clean-up process must never take more than eight hours.

   A. True   **B.** False

128. The first thing that should be done in an uncontrolled hazardous material release is to notify site personnel about the release.

   A. True   **B.** False

129. If victims are in need of medical treatment, they must first be decontaminated.

   A. True   **B.** False

130. The last steps in the follow-up of a HazMat incident is to clean up the area.

   A. True   **B.** False

Unit Five

Strikes, Lockouts, and Labor Relations

131. One of the primary functions of the Protection Officer during a strike is picket-line surveillance.

   A. True   **B.** False

132. Normally, the senior Site Security Supervisor is responsible for all security shift responsibilities.

   A. True   **B.** False

133. The key to good security in labor disputes is apprehension, not prevention.

   A. True   **B.** False

134. An effective company-search program may decrease employee morale.

   A. True   **B.** False

135. A suspension provides the employee with an opportunity to think about the infraction(s) committed and whether the employee wants to continue employment with the company.

   A. True   **B.** False

136. An effective company search program can help a company protect its assets by:

    A. Reducing accident rates

    B. Reducing theft

    C. Reducing the use or possession of contraband on property

    D. All of the above

    E. None of the above

137. Which of the following is *not* considered a type of discipline in labor relations?

    A. A written warning

    B. A suspension

    C. A layoff

    D. A demotion

    E. A discharge

138. Documentation of illegal activities on the picket line could be useful in the following instances:

    A. To support criminal charges

    B. To support company discipline imposed on an employee

    C. To support or defend against unfair labor practice complaints

    D. To support obtaining an injunction

    E. All of the above

139. Because of the serious nature of an employee being discharged for cause, which factor is normally not considered?

    A. Seniority with the company

    B. The age of the employee

    C. The salary or hourly rate of the employee

    D. The marital status of the employee

    E. None of the above

140. The single most-important piece of equipment the Protection Officer should have with them at the picket line is:

   A. A tape recorder

   B. Binoculars

   C. A notebook

   D. A night stick

   E. None of the above

Emergency Planning and Disaster Control

141. In the event of a disaster, the following authorities should become involved as quickly as is practically possible. (Mark the item that is out of place.)

   A. The fire department

   B. The media department

   C. The Red Cross

   D. The police department

142. To ensure effective implementation of a disaster plan, it is important that: (Mark the best answer.)

   A. Each department head has authority to activate the plan

   B. One individual will be responsible

   C. The disaster team is on call at all times

   D. The Chief Executive Officer or that individual's assistant is available

143. It is important to exclude government authorities when developing a local facility disaster-recovery plan.

   A. True   **B.** False

144. The disaster advisory committee should include key personnel from the fire, safety, and security departments, as well as other department personnel.

   A. True   **B.** False

145. When determining which employees should be involved in the development of a disaster advisory committee, the following considerations should be taken into account: (Mark the item that is out of place.)

    A. Knowledge of the facility

    B. After-hours accessibility in the event of an emergency

    C. Ability to exercise good public relations at all times

    D. Length of service with the organization

146. In the event of a disaster, it may be necessary to shut down or limit some facility activities because of: (Mark the item that is out of place.)

    A. The extent of the damage to the facility

    B. The availability of the workforce

    C. The extent and effect of adverse media reports

    D. The availability of internal and external protection units

147. The communications of a warning or alarm must be capable of transmitting throughout the entire facility.

    A. True   **B.** False

148. Security personnel should have access to a current list (including resident telephone numbers) of key individuals and organizations that would be involved in the activation of a disaster plan. This list should include:

    A. Corporation department heads

    B. All employees

    C. The police and fire departments

    D. A and C above

149. Use of the facility public-address system should be limited to use by plant protection personnel in the event of an emergency.

    A. True   **B.** False

150. It is important to determine the number of Security Officers that will be required to ensure that: (Mark the best answer.)

    A. There is sufficient security personnel to direct the police.

    B. There is sufficient plant protection prior to, during, and after the disaster.

    C. Security personnel can fill in for injured workers.

    D. The disaster plan is administered effectively.

V.I.P. Protection, Hostage Conditions

151. Terrorism can be described as:

    A. Decisive and organized

    B. Systematic and theatrical

    C. Planned and calculated

    D. Violent and destructive

152. What quality(s) will "make or break" a PPS more quickly than anything else?

    A. Manners, deportment, and decorum

    B. Education

    C. Determination

    D. Fitness level

153. To truly understand the phenomenon of terrorism, one must have a foundation of:

    A. Knowledge in religion

    B. History

    C. Political science

    D. All of the above

154. Right-wing groups may be described as:

    A. Male or female members

    B. Usually college educated to some degree

    C. Upper class or upper-middle class

    D. None of the above

155. Management of a hostage situation or other crisis event consists of several key elements:

    A. Control

    B. Coordination

    C. Communication and information

    D. All of the above

156. A few things to bear in mind when acting as a security escort include: (Select the incorrect answer.)

    A. Never leave the protectee unguarded

    B. Always be alert and ready to respond to emergencies

    C. Provide protectee itinerary to press

    D. Position yourself between the protectee and possible threats

157. The basic concepts of negotiations are: (Select the incorrect answer.)

    A. Never give hostage takers weapons or intoxicants.

    B. Don't make promises or threats you cannot keep.

    C. Argue if necessary.

    D. Stall for time as much as possible. Say you have to check with your boss.

158. Left-wing groups can be described as: (Select the incorrect answer.)

    A. Communist/socialist

    B. Predominantly male

    C. Young, generally under 45

    D. Upper class or upper-middle class

159. If you are taken hostage:

    A. Be patient

    B. Remain calm

    C. Try to rest

    D. All of the above

160. Animal-rights activists can be described as:

    A. Extensive violence against persons is not used.

    B. No dominant organizational subculture is present within the ranks.

    C. They are not at all easy to spot.

    D. All of the above

Unit Six

Human Relations

161. As a person progresses in the organization, their role tends to diminish in terms of security technical requirements.

    A. True   **B.** False

162. An analysis of human-relations skills required of the Security Officer reveals the following categories that are generic to most situations: (Mark the least-applicable item)

    A. Utilize common sense and discretion

    B. Develop an effective dialogue

    C. Work as a team member and be sensitive to others

    D. Be involved in professionalism and the learning process

163. Love and Belonging is the highest order of Maslow's Hierarchy of Needs.

    A. True   **B.** False

164. The Physiological aspect of Maslow's Hierarchy of Needs (lowest order) includes: (Mark the incorrect item.)

    A. Food and drink

    B. Shelter

    C. Sleep and sex

    D. Safety and security

165. The authoritative style of leadership functions best when dealing with a mature kind of individual.

    A. True   **B.** False

166. The methods best employed when implementing a participative, democratic style of leadership are: (Mark the two best answers.)

    A. Persuasion

    B. Psychological support

    C. Inspiration, motivation

    D. Authority, discipline

167. One of the premises of Transactional Analysis is that each person decides on their own life plan and they are not responsible for maintaining or changing it.

    A. True   **B.** False

168. Identify each of the Ego State Reactions to the following situation, for example, Parent (P), Adult (A), Child (C). Mark the answer key with the appropriate abbreviation: (P), (A), or (C).

    A. "A fellow Security Officer gets an unexpected promotion."

    B. "Well, she certainly deserved it. After all, with all those children to feed, she will need all the help she can get."

    C. "Bad news—she got the promotion because she always 'sucks-up' to the Chief."

    D. "I thought I was more qualified. I guess I have not given her enough credit."

169. The life position, I'm O.K., You're Not O.K., results when an infant has cold, nonstroking parents; who feel defeated and could be suicidal.

    A. True   **B.** False

170. Psychologists have identified a number of barriers to effective communications: (Mark the correct answer.)

    A. Language, speech difficulties, or an inability to convert the message into meaningful dialogue

    B. Distractions because of panic, danger, emotionalism, or rage

    C. Personal bias, prejudice, or status

    D. All of the above

Interviewing Techniques

171. If the Protection Officer is unable to control the interview:

    A. Time will be lost

    B. Facts may be forgotten

    C. The psychological advantage shifts

    D. All of the above

172. An experienced interviewer need not have a game plan when interviewing.

    A. True   **B.** False

173. The success or failure of an interview depends entirely on the skill of the investigator.

    A. True   **B.** False

174. The location of the interview should be chosen by the subject, so they will be more at ease in familiar surroundings.

    A. True   **B.** False

175. In most instances, the content of an incident will be covered in more than one conversation.

    A. True   **B.** False

176. Rapid-fire questioning techniques will most likely:

    A. Cause the subject to tell the truth

    B. Confuse the subject

    C. Cause the subject to lie

    D. B and C above

177. By asking a series of questions early in the interview, you:

    A. Condition the subject to believe that if you want information, you will ask

    B. Lead the subject to believe that everything they tell you has significance

    C. Cause the subject to withhold information by putting them on guard

    D. All of the above

178. Obstacles to conversations in an interview are:

    A. Specific questions

    B. Yes and no questions

    C. The use of leading questions

    D. All of the above

179. While getting acquainted with a subject, you should:

    A. Be opinionated

    B. Put the subject at ease

    C. Be informal

    D. B and C above

180. Interviewees can be afraid of:

   A. Incriminating themselves

   B. Becoming a witness

   C. The "uniform"

   D. All of the above

Stress Management

181. Feelings of defeat, frustration, and/or depression activate the adrenal cortex.

   A. True   **B.** False

182. A communications system of nerve cells throughout the body is the sympathetic nervous system.

   A. True   **B.** False

183. Adrenalin and epinephrine are the same hormones.

   A. True   **B.** False

184. During stress, the blood thins significantly.

   A. True   **B.** False

185. Sustained and repeated alarms may cause:

   A. Heart problems

   B. Gastrointestinal problems

   C. Susceptibility to disease

   D. Marriage problems

   E. All of the above

186. The burnout process is:

   A. An excuse for sloppy work

   B. A socially learned "cop-out"

   C. A psychiatric weakness

   D. A recognized stress syndrome

   E. A common stress reaction

187. Lazarus found a high relationship among depression, exhaustion, anxiety, and illness in:

  A. The number of friends of the opposite sex

  B. The size of one's bank balance

  C. The number of children in the family

  D. The amount of overtime worked

  E. The number of everyday life hassles

188. Effective coping mechanisms are based on:

  A. Time management

  B. Exercise and sleep

  C. Values and attitudes

  D. Intimacy and family life

  E. All of the above

189. Relaxation techniques are:

  A. Sex drives

  B. A natural tranquilizer

  C. Evil forces

  D. Inhibitions

  E. All of the above

190. Each of us has spent a lifetime developing:

  A. Values

  B. Skills

  C. Destructive behaviors

  D. Attitudes

  E. All of the above

Crisis Intervention

191. Crisis-intervention techniques are designed to provide more control of the eventual outcome of a crisis incident.

  A. True   **B.** False

192. "Proxemics" refers to how we deliver our words or verbally intervene.

    A. True   **B.** False

193. The objective of any crisis-development situation is to defuse the situation while maintaining the safety and welfare of all involved.

    A. True   **B.** False

194. Anger and frustration are a normal reaction of the Protection Officer after a crisis situation.

    A. True   **B.** False

195. It is better to handle a crisis situation one-on-one; a group would only increase tension.

    A. True   **B.** False

196. During a crisis-development situation, there are four distinct and identifiable behavior levels. The list does not include:

    A. Anxiety

    B. Defensiveness

    C. Disinterest

    D. Anger/frustration

197. During the evaluation stage of the management of disruptive behavior, the first thing the Protection Officer must do is:

    A. Physically restrain the individual

    B. Clear the area of spectators

    C. Implement an action plan

    D. Find out what is going on

198. Personal space is generally defined as:

    A. 1.5 to 3 feet from the individual

    B. The area dictated by the individual

    C. The room occupied by the individual

    D. An arm's length away

199. Paraverbal communication does not deal with one of the following, which is:

   A. Tone

   B. Movement

   C. Rate

   D. Volume

200. When using the team approach to intervention, there is a maximum number of team members, which is:

   A. 1

   B. 5

   C. 2

   D. 4

Unit Seven

Security Awareness

201. Deterring a potential culprit from violating a facility can best be achieved by: (Mark the incorrect item.)

   A. Creating the impression that there may be a high degree of security

   B. Posting signs that indicate security measures are present, even if they are not in place

   C. Insuring a strong presence of security personnel in uniform exists at all times

   D. Making all employees and those with access to a facility aware of the effective security system

202. The visibility of a security program is advisable because:

   A. It tends to frighten employees

   B. It provides a strong form of deterrence

   C. Visitors to the facility may tell their law-breaking friends

   D. It may well create a better atmosphere of trust

203. There need not be any physical evidence of an effective security program to develop security awareness among employees.

   A. True   **B.** False

204. Onsite physical security components that could be utilized to secure an eight-story building could be: (Mark the item that is out of place.)

    A. A fence

    B. A security consultant

    C. An uniformed Security Officer

    D. CCTV

205. An effective security program could require that individuals leaving a privately owned facility allow security personnel to examine the interior of their attaché cases.

    A. True   **B.** False

206. Increased security measures equate to the likelihood of reduced criminal activity.

    A. True   **B.** False

207. If a culprit is caught committing a criminal offense on or in relation to the facility, the matter should be: (Mark the best answer.)

    A. Concealed from other employees

    B. Published in the company newsletter

    C. Made known to all employees

    D. Brought to the attention of the guard

208. The security department should make its success discreetly known throughout the facility.

    A. True   **B.** False

209. Prevention is the bottom line to an effective security-awareness program.

    A. True   **B.** False

210. Organizational security awareness is best achieved by:

    A. Effective access control

    B. Imposing punishment when necessary

    C. Integrated security systems

    D. Enlightened and informed employees

Security Investigations

211. The first priority when responding to a crime scene is to:

   A. Detain the perpetrator if they are still present

   B. Attend to injured person(s)

   C. Preserve physical evidence

   D. Complete a preliminary report

212. The follow-up investigation begins with:

   A. An examination of the data provided in the preliminary investigation

   B. An assessment of the crime/accident scene

   C. Initial interviewing of witnesses

   D. Follow-up interviews of witnesses

213. Investigation is best defined as:

   A. A subjective method used to prepare cases for prosecution

   B. A subjective process used to discover facts

   C. An objective process used to discover facts

   D. An objective method of reporting, so cases may be prosecuted

214. Auditing is best defined as:

   A. An activity for accountants to be engaged in

   B. A supervisory evaluation of conditions

   C. A systematic method of examining financial records

   D. A check or investigation as to whether operations are proceeding as expected

215. Which of the following is *not* an acceptable audit practice?

   A. Interviewing personnel about procedures

   B. Advising personnel about the results of the audit only when they are subject to disciplinary action

   C. Conducting drills

   D. Observing job behavior or systems

216. Which of the following is an acceptable technique to use during interrogations?

    A. Pointing out inconsistencies in statements

    B. Promising the subject considerations if they cooperate

    C. Promising to obtain counsel for the subject if they confess

    D. Placing the subject in a room surrounded by large, armed, and oppressive security personnel

217. The success of the follow-up investigation is directly dependent on:

    A. The availability of witnesses

    B. The availability of physical evidence

    C. The preliminary investigative effort

    D. The technological resources available to the investigator

218. While behavior analysis can be used to investigate almost any type of incident, it is most commonly used in the investigation of:

    A. Credit card fraud

    B. Robbery

    C. Bad checks

    D. Embezzlement

219. An important concept in managing investigations is to:

    A. Appropriate budgets

    B. Discipline investigators

    C. Guide the investigation toward the conviction of the defendant

    D. Establish objectives

220. When testifying in court, a Security Officer should:

    A. Answer questions with a "yes" or "no"

    B. Provide extensive detail on all questions

    C. Begin answering questions with "I think"

    D. Present a positive image, never admitting that "I don't know"

Managing Employee Honesty

221. Which of the following employees steal?

    A. Managers

    B. Supervisors

    C. Line employees

    D. B and C only

    E. A, B, and C

222. Waste containers are favorite stash places for employees who steal.

    A. True   **B.** False

223. The first step in employee-theft prevention is to learn what can be stolen.

    A. True   **B.** False

224. Protection Officers who miss an assigned round are examples of which WAECUP threat?

    A. Waste

    B. Error

    C. Crime

225. It is better to catch employees who are thieves than to reduce the opportunity for theft.

    A. True   **B.** False

226. It is important to know who is authorized to take trash outside.

    A. True   **B.** False

227. Employee thieves remove company property:

    A. In their own vehicles

    B. In company vehicles

    C. By walking out with it

    D. All of the above

228. First observe, and then report.

   A. True   **B.** False

229. When in doubt about the search policy, ask:

   A. Any supervisor

   B. Any employee

   C. Your supervisor

   D. The employee you want to search

230. A change in the workplace environment that is most likely to raise the potential for employee theft is:

   A. Your vacation

   B. The arrival of new valuable items

   C. New office setup (of desks, and so forth)

   D. Your new uniform

Substance Abuse

231. Substance abusers only steal from their employer.

   A. True   **B.** False

232. "Crack" is the smoked form of:

   A. Marijuana

   B. Methamphetamine

   C. Valium

   D. Cocaine

233. Marijuana users smoke "sustained low dosages" while at work to avoid detection.

   A. True   **B.** False

234. Which is an investigative technique used to detect drug dealing on the job?

   A. Undercover investigator

   B. Hidden cameras

   C. Employee interviews

   D. All of the above

235. Workplace drug dealers generally sell their drugs in bathrooms, parking lots, vehicles, and secluded areas.

     A. True   **B.** False

236. LSD and heroin fall under the same drug category.

     A. True   **B.** False

237. Drug-abuse prevention is the total responsibility of management.

     A. True   **B.** False

238. One of the key individuals in a counter-drug-abuse program is the line supervisor.

     A. True   **B.** False

239. Which drug does not fall under the narcotic drug category?

     A. Heroin

     B. Codeine

     C. LSD

     D. Opium

240. The risk of acquiring serum hepatitis occurs only when snorting cocaine through a metal tube.

     A. True   **B.** False

Unit Eight

Legal Aspects

241. The purpose of our legal system is to:

     A. Set down our obligations to each other

     B. Set penalties for breaching these obligations

     C. Establish procedures to enforce those obligations

     D. All of the above

242. The common law never changes.

     A. True   **B.** False

243. The doctrine of case law states that a court must stand by previous decisions.

    A. True   **B.** False

244. Statutes are changed:

    A. Never

    B. To fill a need in our society

    C. Only by a level of government higher than the one that passed the law

    D. Whenever there is a change in government

245. The prosecutor's job is to get compensation for the victim.

    A. True   **B.** False

246. The police will investigate:

    A. Civil matters

    B. Criminal matters

    C. Whatever they are paid to investigate

    D. All of the above

247. A civil action may not commence until the criminal courts are finished.

    A. True   **B.** False

248. A warrant to arrest may be executed by:

    A. A private citizen

    B. The police

    C. A Security Officer

    D. Anyone who apprehends the suspect

249. A Security Officer can arrest anyone they find committing an offense on property they are protecting.

    A. True   **B.** False

250. Confessions cannot be admitted in court as evidence if:

  A. The accused later denies guilt

  B. It was not in writing

  C. It was not signed by the accused

  D. It was not voluntary

Unit Nine

Physical Fitness, Exercise Program

251. Muscular endurance is defined as the maximum tension a muscle can exert when contracted.

  A. True   **B.** False

252. The F.I.T.T. principles:

  A. Stands for frequency, intensity, target, and type

  B. Describes the guidelines for setting up an exercise

  C. Should only be used as prescribed by a qualified fitness appraiser

  D. Is only applicable to the cardiovascular component of fitness

253. An aerobic test can be used to determine cardiovascular fitness.

  A. True   **B.** False

254. Flexibility exercises:

  A. Should be integrated into the warm-up and cool-down phases of an exercise program

  B. Are most effective if ballistic and bobbing techniques are used

  C. Can help prevent muscle injury and soreness

  D. Both A and C are correct

  E. All of the above

255. Exercising two times per week is the minimum requirement for improving fitness.

  A. True   **B.** False

256. Working above the target heart-rate zone is recommended for achieving a higher fitness level.

  A. True   **B.** False

257. Physical fitness is an important component of health.

   A. True   **B.** False

258. Physical activities that can improve cardiovascular fitness:

   A. Are repetitive in nature, like calisthenics

   B. Should hurt, because that ensures the appropriate intensity level

   C. Should be fun and personal

   D. Require keeping the heart rate between 160 and 180 beats per minute

   E. None of the above

259. A proper warmup should:

   A. Get your heart rate in the target zone

   B. Increase the blood flow to the muscles that will be working

   C. Be about 75 percent of your maximum heart rate

   D. All of the above

   E. None of the above

260. The best exercises for a weight-loss program:

   A. Are repetitive and continuous in nature—dynamic

   B. Are low in intensity and long in duration

   C. Do not use gimmicks, but encourage a lifestyle change

   D. Are not necessarily aerobic in nature

   E. All of the above

Unit Ten

First Aid

261. Putting on splints and bandages are treatments done during the first action.

   A. True   **B.** False

262. A clear airway must be established before artificial respiration can be started.

   A. True   **B.** False

263. When doing CPR, it is only necessary to manually circulate the blood for the person who has heart stoppage.

     A. True   **B.** False

264. Brain damage will normally occur within four to six minutes in a person who has had heart stoppage.

     A. True   **B.** False

265. Two important treatments for shock are keeping the injured person warm and treating the injuries.

     A. True   **B.** False

266. When you find an injured person, you want to do first things first. After insuring there are no hazards to you or the injured person, you would then:

     A. Check the casualty for bleeding

     B. Check the casualty's level of consciousness

     C. Check the casualty's limbs for deformity

     D. Check the casualty for possible shock

267. Joints are held tightly in their sockets by:

     A. Ligaments

     B. Cartilage

     C. Muscles

     D. Tendons

268. The proper care for a person with a sucking chest wound includes to:

     A. Loosely bandage a sterile dressing over the wound

     B. Snugly bandage a sterile dressing over the wound

     C. Pack a sterile dressing into the wound

     D. Seal the wound with an airtight dressing

269. If a person has fainted, the aim of a first-aid giver is to:

    A. Immediately place the person in a sitting position, so they can breathe easier

    B. Lay the person down and elevate the lower extremities (legs), if possible

    C. Elevate the head and shoulders to restrict the blood flow to the brain

    D. Give sips of water to build up the body fluids

270. You have a casualty with an area of their arm burned. The skin appears red, there are a few blisters, and the person is experiencing pain. This burn is classified as:

    A. 1st degree

    B. 2nd degree

    C. 3rd degree

    D. 4th degree

Unit Eleven

Use of Force

271. A subject may admit wrongdoing during what period?

    A. Debriefing

    B. Physical confrontation

    C. Assault

    D. Reprimand

272. Calmness is:

    A. Contrasting

    B. Contagious

    C. Lethal

    D. Tactical

273. The definition of "reasonable use of force" is the amount of force:

    A. Necessary to hurt an aggressor

    B. Needed to incapacitate an aggressor

    C. Equal to or slightly greater than the force of an aggressor

    D. Necessary to seriously scare an aggressor

274. When writing a report, in the use of force cases, it is appropriate for all of the following, except:

   A. The report follows a chronology

   B. The report has contradictions

   C. The times stated match other reports and records

   D. The facts stated match other reports and records

275. Which of the following is a recommended formula for self-control?

   A. Intellect/Emotions = Control

   B. Emotions/Intellect = Control

   C. Training/Intellect = Control

   D. Problem/Emotions = Control

276. "Tachy-psyche effect" can be identified from:

   A. Proper background investigation

   B. Physiological characteristics

   C. Family background

   D. Ethnic or racial make-up

277. When approaching apparently rational subjects, officers should do all of the following, except:

   A. Respect the subjects' dignity

   B. Become part of the problem

   C. Shout commands

   D. Be aggressive

278. What is the cornerstone of officer survival?

   A. The use of force

   B. Crisis intervention

   C. Conflict resolution

   D. Aggressive behavior

279. Following the order of the confrontation continuum, the first step is officer presence, and then initial communications. The next step is:

   A. Soft, empty hand control

   B. Verbal commands

   C. Chemical incapacitation

   D. Empty hand impact

280. When dealing with aggression, which of the following is good advice?

   A. Get or give space if the encounter gets tough

   B. Avoid humiliating subjects

   C. Recognize your emotions

   D. All of the above is good advice

Unit Twelve

Public Relations

281. Security officers develop the "malcontent syndrome" by working shorter hours, having days off, and working on good assignments.

   A. True   **B.** False

282. To be prepared for the unexpected, Security Officers should always carry a flashlight, a watch, a notepad, and:

   A. A screwdriver

   B. A pocket knife

   C. A two-way radio, ready in hand

   D. Identification, hanging on their neck

283. To show interest when greeting a customer, a great opening statement would be:

   A. May I help you?

   B. I'll be there in one sec, hang on.

   C. Hi! What can I get for you?

   D. How may I help you?

284. The successful Protection Officer will know every other department's products and services.

   A. True   **B.** False

285. Which of the following selections is an in-depth evaluation of identified threats, probability hypotheses, vulnerability studies, and security surveys of facilities and systems?

    A. Media relations

    B. Supervisor analysis

    C. Training reports

    D. Risk analysis

286. Whenever possible:

    A. Stay out of other people's business

    B. Let the supervisor handle the situation

    C. Be polite, but distant

    D. Help others as much as possible

287. Present a professional appearance:

    A. When meeting the press

    B. Only during inspections

    C. Only when you have to

    D. At all times

288. Security personnel must be:

    A. Managers

    B. Teachers

    C. Salespeople

    D. Instructors

289. Each contact with a person should be treated as:

    A. A procedure

    B. A promotion tool

    C. A client

    D. A process

290. The security department alone is capable of carrying out a successful public-relations program.

    A. True   **B.** False

Police and Security Liaison

291. Relationships between law enforcement and private security in the early 1980s were rated fair to good.

A. True   **B.** False

292. In 1976, the Private Security Advisory Council, through the U.S. Department of Justice, identified two main factors that contributed to poor relationships between law enforcement and private security. Identify the two correct factors.

A. Their inability to clarify role definitions

B. They often practiced stereotyping.

C. Lack of funding

D. Resentment

293. Liaison plays a less-than-substantial role in our daily functions as security professionals.

A. True   **B.** False

294. By the year 2000, Security Officers will outnumber Law-Enforcement Officers three to one.

A. True   **B.** False

295. The cost of economic crime in 1990 was estimated at:

A. $110 million

B. $120 million

C. $1 billion

D. $114 billion

296. According to the Hallcrest Report II, Private Security Trends (1970 to 2000), private security is America's secondary protective resource in terms of spending and employment.

A. True   **B.** False

297. There appears to be a growing potential for contracting private security to perform the following activities: (Select the incorrect activity.)

A. Courtroom security

B. Special-event security

C. Hostage negotiations

D. Traffic control

298. The key to maintaining a good working relationship with law enforcement personnel is to: (Select the best answer.)

    A. Maintain a good public perception

    B. Enroll in law-enforcement programs

    C. Send them all of your reports

    D. Maintain a high-level of physical fitness

299. In our society, we are judged by how we look.

    A. True   **B.** False

300. Common ground for interaction between law enforcement and private security is crime.

    A. True   **B.** False

Ethics and Professionalism

301. Bottom and Kostanoski's WAECUP model asserts that losses stem from all of the following, except:

    A. Waste

    B. Accident

    C. Crime

    D. Professional practices

302. Which is not in the International Foundation for Protection Officers' Code of Ethics?

    A. Respond to the employer's professional needs

    B. Respond to the employer's personal needs

    C. Protect confidential information

    D. Dress to create professionalism

303. Which of the following is not true?

    A. The *P* in "professional" is for precise.

    B. The *E* in "professional" is for ethical.

    C. The *I* in "professional" is for informed.

    D. The *A* in "professional" is for accompany.

304. Professionals always think in terms of their:

    A. Looks

    B. Duties

    C. Authority

    D. Image

305. The acronym PORT includes all, except:

    A. Personal

    B. Options

    C. Responsibilities

    D. Time

306. The letter *R* in the acronym PORT stands for:

    A. Review

    B. Responsibility

    C. Respond

    D. None of the above

307. Deportment can be defined as:

    A. Making a poor decision

    B. A continuous process

    C. Writing out the PORT

    D. How one carries oneself

308. Manners that promote a professional image do not include:

    A. Standing to greet people

    B. Praising others when appropriate

    C. Allowing people to talk without interruption

    D. Tucking a napkin into your belt

309. Unethical behavior may occur because of:

    A. Fatigue

    B. Quality recruitment

    C. Proper and adequate training in decision-making

    D. Quality supervision where superiors set a good example

310. A true professional has all the following characteristics, except:

    A. Education related to the profession

    B. A consistent record of winning court cases

    C. A commitment to the profession

    D. Experience within the profession

Exam 2

The following test is based on the International Foundation for Protection Officers program. If you earn a passing score, you should be able to pass other security certification and licensing tests. Although the International Foundation for Protection Officers doesn't endorse the Registered Security Professional, you can earn the certification by passing the following test and submitting copies of the proper documents that show you are a security professional.

To receive your certificate, submit your answers to the following exam (a passing score is 70 percent). An answer sheet appears on page 280. Enclose copies of any two of the following (don't send originals because they will not be returned).

- City of state locksmithing or alarm technician license
- Drivers license or state issued identification
- Locksmith or alarm system suppliers invoice
- Certificate from locksmithing, alarm system, security officer, or other security school program
- Yellow pages listing
- Association membership card or certificate
- Locksmith, alarm technician, or safe technician bond card or certificate
- Letter from your employer or supervisor on company letterhead stating that you work as a locksmith, security officer, alarm system technician, safe technician, or toher security professional
- Letter of recommendation from a Registered Professional Locksmith or Registered Security Professional
- Copy of an article you have had published in a locksmithing, alarm system technician, or other security-related trade journal
- ISBN number and title of a security-related book you have written

Unit One

The Evolution of Asset Protection and Security

1. What year was the Black Tom Island explosion?

    A. 1916

    B. 1948

    C. 1966

    D. 1912

2. A light cavalry unit established during the Texan Revolution would be called:

    A. Mexican Calvary

    B. Texas Rangers

    C. Pennsylvania Police

    D. National Guard

3. Allan Pinkerton established _____ and forbade the acceptance of awards from his men.

    A. Law guidelines

    B. A reward

    C. A code of ethics

    D. New uniforms

4. In the later nineteenth century, the American West had its share of security problems in the form of outlaws and _____.

    A. Hostile Indians

    B. Slaves

    C. Bankers

    D. Rangers

5. In what year was The National Burglar and Fire Alarm Association formed?

    A. 1871

    B. 1667

    C. 1948

    D. 1666

6. In what year was the Peshtigo Fire in Wisconsin?

   A. 1871

   B. 1666

   C. 1631

   D. 1948

7. In 1974 the _____ was passed, establishing security programs at airports.

   A. Air Rage Act

   B. Antihijacking and Air Transportation Act

   C. Communication Act

   D. Loss Prevention Act

8. _____ Theft incorporates theft, repackaging, and distribution of the stolen product.

   A. Organized Retail

   B. Loss Prevention

   C. Retail Security

   D. Resource Management

Unit Two

Field Notes and Report Writing

9. At some point, an officer may be requested to show their notebook to::

   A. Their previous supervisor

   B. The courts

   C. A competitor

   D. All of the above

   E. None of the above

10. Which of the following should an officer have readily available when writing reports?

    A. A cup of coffee

    B. Extra forms

    C. A dictionary

    D. All of the above

    E. None of the above

11. Security reports could be viewed by:

     A. A judge

     B. A defense lawyer

     C. A security manager

     D. All of the above

     E. None of the above

12. After reading your daily reports, which individual or group would most likely benefit from the contents?

     A. The public

     B. The private justice system

     C. Your fellow officers

     D. All of the above

     E. None of the above

13. When writing an accurate report, it is imperative to refer to:

     A. A phone book

     B. Your notes

     C. Previous reports

     D. All of the above

     E. None of the above

14. Proper notes are the first step in forming a permanent record of events as they occurred.

     A. True   **B.** False

15. Notes are not an essential part of proper report preparation.

     A. True   **B.** False

16. The officer's ability to patrol is judged solely on the basis of their notebook.

     A. True   **B.** False

Observation Skills and Memory

17. The smaller the object, the farther away the observer will be able to recognize it.

    A. True   **B.** False

18. Under normal conditions of visibility, a person with distinctive features can be recognized by friends and relatives at:

    A. 75 yards

    B. 50 yards

    C. 100 yards

    D. 125 yards

19. Which substance may temporarily kill your sense of smell?

    A. Gunpowder

    B. Wood smoke

    C. Ether

    D. Electric smoke

20. When we use our senses effectively, we are thinking and being aware.

    A. True   **B.** False

21. Each person's ability to recall information from memory is the same, regardless of the amount of practice.

    A. True   **B.** False

22. When patrolling, you should stop occasionally just to listen.

    A. True   **B.** False

23. To help improve your sight, you should:

    A. Be aware of what you look at

    B. Insure your vision is in peak condition

    C. Insure you understand the factors that affect your vision

    D. All of the above

24. The position of the observer in relation to the subject can alter the observer's perception of the subject.

   A. True   **B.** False

Patrol Principles

25. Which of the following can enhance the patrolman's skill and ability? (Select the best answer.)

   A. Proper training

   B. Preparation for patrol

   C. Professional work habits

   D. All of the above

26. Security Officers should act and look professional only while at work.

   A. True   **B.** False

27. Detecting criminals is the major responsibility of Security Officers.

   A. True   **B.** False

28. Under no circumstance should an officer leave their post until properly relieved.

   A. True   **B.** False

29. It is not essential that the Security Officer document all observations.

   A. True   **B.** False

30. Based on organizational needs, there are several major purposes of patrol. (Select the best answer.)

   A. Respond to emergencies

   B. Prevention and deterrence of crime

   C. Detection of criminal activity

   D. All of the above

31. The function of security is to prevent and control loss.

   A. True   **B.** False

32. The WAECUP Theory includes which of the following?

    A. Waste

    B. Error

    C. Accident

    D. Crime

    E. All of the above

Safety and the Protection Officer

33. Which of the following safety/security conditions should be of concern to the Protection Officer on patrol? (Mark the out-of-place entry.)

    A. Boxes blocking an exit

    B. A flaw in computer interfacing

    C. An obstructed CCTV camera lenses

    D. An overheating electrical motor

34. Unsafe conditions, such as poor housekeeping, can be a major contributor to an accident.

    A. True   **B.** False

35. Accident prevention measures taken by the Protection Officer while on patrol relate to: (Select the best answer.)

    A. Taking an active role in Safety Committee meetings

    B. Monitoring employee behavior

    C. Noting and reporting safety hazards

    D. Providing meaningful reports that can be interpreted by top management

36. Employee training has a positive effect in terms of developing on-the-job safety practices.

    A. True   **B.** False

37. Providing direct employee-safety training is the responsibility of: (Select the best answer.)

    A. Manager

    B. Supervisor

    C. Safety Committee

    D. Protection Officer

38. Safety meetings are exclusively for the Safety Committee.

  A. True   **B.** False

39. Safety contests with awards are a good way to improve safety practices by employees because:

  A. This practice resembles a lottery.

  B. This practice gets employees involved with managers.

  C. This practice increases employee safety awareness and motivation.

  D. This practice is proven effective in strengthening union-management relations.

40. Safety posters have been proven as the best method of conveying employee safety-awareness practices.

  A. True   **B.** False

Traffic Control

41. A corner position provides a better view of traffic than a center-of-the-intersection position.

  A. True   **B.** False

42. When traffic is congested and motorists desire frequent turns that result in slowing the flow of traffic, they should be:

  A. Stopped

  B. Obliged

  C. Forced to go straight through

  D. Pulled over to give the intersection a chance to clear

43. Proper officer protection against the elements is an important factor in maintaining efficient traffic control.

  A. True   **B.** False

44. Prompt compliance to hand signals is dependent on the officer's ability to:

  A. Use uniform, clearly defined hand signals

  B. Quickly assess traffic flow needs

  C. Quickly assess congestion problems

  D. All of the above

45. When an emergency vehicle is approaching, you stop:

    A. All pedestrian traffic

    B. Only vehicles on the street on which the emergency vehicle is approaching

    C. The emergency vehicle

    D. All vehicular and pedestrian traffic

46. When directing traffic, priority of movement is determined by the amount of traffic flow in each direction.

    A. True   **B.** False

47. When attempting to attract a motorist's attention with a whistle, give:

    A. One long blast

    B. Two long blasts

    C. Two short blasts

    D. One short blast

48. The property owner of a private parking lot is responsible for controlling traffic and patrolling the area.

    A. True   **B.** False

Crowd Control

49. A crowd may initially exist as a casual or temporary assembly having no cohesiveness.

    A. True   **B.** False

50. It is important for the Protection Officer to be able to quickly determine if a gathering may become uncontrollable.

    A. True   **B.** False

51. Members of a crowd that assemble for a sporting event depend on each other for support and have a unity of purpose.

    A. True   **B.** False

52. Persons joining a crowd tend to accept the ideas of its dominant members without realization or conscious objection.

    A. True   **B.** False

53. An individual can lose self-consciousness and identity in a crowd.

    A. True   **B.** False

54. There is potential for mass discord whenever people gather at:

    A. Athletic events

    B. Parades

    C. Protest rallies

    D. All of the above

    E. None of the above

55. In dealing with persons in a crowd situation, a Protection Officer should:

    A. Use good judgment and discretion

    B. Remain impartial and courteous

    C. Refrain from derogatory remarks

    D. All of the above

    E. None of the above

56. Emotional reactions resulting in the formation of unruly crowds are often associated with:

    A. Absence of authority

    B. Religious/racial differences

    C. Economic conditions

    D. All of the above

    E. None of the above

Unit Three

Physical Security Applications

57. The final step in the process of conducting a security vulnerability analysis is the identification of assets.

    A. True   **B.** False

58. Security lighting is designed solely to detect intruders.

    A. True   **B.** False

59. Wired glass is considered to be burglar- and vandal-resistant.

    A. True   **B.** False

60. Ultrasonic detectors are *not* motion sensors.

    A. True   **B.** False

61. Ultrasonic detectors are not recommended for use in an area that may be subjected to air turbulence.

    A. True   **B.** False

62. The optical coded badge can be recognized by:

    A. Its shape

    B. The solid black bar across the back of the card

    C. The holes punched in the card

    D. The use of the "bar" code

63. Closed-circuit television sequential switchers are highly recommended because:

    A. They stamp the date and time on each video tape.

    B. They allow for pan-and-tilt control of cameras.

    C. They allow one monitor to be used with several cameras.

    D. They allow an operator to zoom in for a close-up.

64. The minimum recommended gauge for chain-link fence fabric is:

    A. 6 gauge

    B. 14 gauge

    C. 9 gauge

    D. 12 gauge

Alarm System Fundamentals

65. What type of alarm monitoring occurs when it is accomplished by your own employees from your company's own security-control center?

    A. Central station

    B. Local monitoring

    C. Proprietary monitoring

    D. Networked surveillance

66. Buttons used for duress alarms should be obvious and in plain view of everyone.

    A. True   **B.** False

67. Area protection is also sometimes called:

    A. Volumetric protection

    B. Spot protection

    C. Perimeter protection

    D. Medium security

68. As a general rule, area sensors come in two varieties. These are:

    A. Analog and digital

    B. Passive and active

    C. Internal and external

    D. Integrated and digital

69. BMS stands for:

    A. Basic Monitoring System

    B. Balanced Magnetic Switch

    C. Barometric Measuring Standard

    D. Balanced Monitoring System

70. These motion sensors detect a change in the thermal-energy pattern caused by a moving intruder and they initiate an alarm when the change in energy satisfies the detector's alarm criteria.

    A. Microwave

    B. Passive InfraRed (PIR)

    C. Dual-technology sensors

    D. Pressure mats

71. Dual-technology sensors combine two different technologies in one unit.

    A. True   **B.** False

72. Fixed-duress devices are mechanical switches permanently mounted in an inconspicuous location.

    A. True   **B.** False

Central Alarm Stations and Dispatch Centers

73. Name three types of alarms that are received in the control room.

    A. Fire, Burglar, and Auxiliary

    B. Fire, Panic, and Theft

    C. Censor, Fire, and Theft

    D. Auxiliary, Fire, and Panic

74. Only some communications are relayed and recorded by the dispatcher.

    A. True   **B.** False

75. The security control room operates only during the day.

    A. True   **B.** False

76. What device is used by the dispatcher when playing back or clarifying a call?

    A. Digital audio recorder

    B. Video recorder

    C. Tape recorder

    D. CD player

77. Long phone cords or headsets prevent neck strain.

    A. True   **B.** False

78. What device controls cameras and allows pan, tilt, and zoom?

    A. Power box

    B. CCTV controller

    C. Remote controller

    D. Digital signal switch

79. Digital radio-base stations are user-friendly.

    A. True   **B.** False

80. The computer-assisted dispatch program does not allow the operator to enter comments and call details.

    A. True   **B.** False

Access Control

81. When left with the decision whether to allow access, in most cases, allowing access is best.

    A. True   **B.** False

82. A restricted keyway combined with strict key control is an effective way of controlling access.

    A. True   **B.** False

83. _____ recognition technology is integrated in CCTV systems to identify individuals who are wanted by law enforcement.

    A. Keyway

    B. Card

    C. Facial

    D. Hand

84. All key-control documentation should be considered public information and should be posted at the front desk for public viewing.

    A. True   **B.** False

85. Installation of an electronic-access system will not significantly enhance the control of keys.

    A. True   **B.** False

86. The use of _____ and watermarking make the replication of company identification more difficult.

    A. Dye sublimation

    B. Holograms

    C. Wiegand

    D. Locks

87. Dallas touch memory is by far the most popular access-card technology.

    A. True   **B.** False

88. The document used to sign in and out at the front entrance of a facility is often called a register.

    A. True   **B.** False

Unit Four

Introduction to Computer Security

89. An e-mail might appear to come from a company executive directing the sale of company assets. In fact, the e-mail could have originated from someone completely outside the organization. This is an example of:

    A. Impersonation

    B. Eavesdropping

    C. Hacking

    D. Social engineering

90. The two major problems with passwords are when they are easy to guess based on knowledge of the user (for example, a favorite sports team) and when they are:

    A. Vulnerable to dictionary attacks

    B. Shared by users

    C. Created by the system manager

    D. Based on a foreign language

91. A *back door* is a potential weakness intentionally left in the security of a computer system or its software by its designers.

    A. True   **B.** False

92. According to the text, motivations for attacks can include all of the following, except:

    A. Committing information theft and fraud

    B. Disrupting normal business operations

    C. Impressing members of the opposite sex

    D. Deleting and altering information

93. In general, web sites are open doors that often invite an attack.

    A. True   **B.** False

94. Plans for backing up computer files should include:

 A. Regularly scheduled backups

 B. Types of backups

 C. The information to be backed up

 D. All of the above

95. In this tactic, a computer system presents itself to the network as though it were a different system.

 A. E-mail tracking

 B. Network spoofing

 C. Encrypting

 D. Social engineering

96. This technique involves tricking people into revealing their passwords or some form of security information.

 A. Network spoofing

 B. Data mining

 C. Social engineering

 D. Password cracking

Information Security

97. Budget proposals are not critical information, as they eventually do become public knowledge.

 A. True   **B.** False

98. A *trade secret* is an agreement to privately exchange products between two companies.

 A. True   **B.** False

99. Physical security concepts aid in the protection of information.

 A. True   **B.** False

100. Hackers are a type of threat to sensitive information.

 A. True   **B.** False

101. Threats to information could include:

    A. Being copied

    B. Being stolen

    C. Being destroyed

    D. All of the above

102. One of the key points in determining how to protect information is:

    A. The day of the week

    B. The name of the document

    C. The form of the information itself

    D. None of the above

103. Paper documents are often considered easier to secure, as they:

    A. Are not dependent on power

    B. Are less valuable

    C. Do not deteriorate with age

    D. All of the above

104. Which types of communications components should be inspected?

    A. Phone lines

    B. Fax lines

    C. Data lines

    D. All types should be inspected

Unit Five

Explosive Devices, Bomb Threats, and Search Procedures

105. Explosives may be made, bought, or stolen.

    A. True   **B.** False

106. Most bomb threats are eventually found out to be hoaxes.

    A. True   **B.** False

107. The Bomb Threat Checklist should not be distributed, until a threatening call is identified.

    A. True   **B.** False

108. The use of two-way radios is discouraged when performing a bomb search.

    A. True   **B.** False

109. Information to be sought, when handling a bomb-threat call, might include:

    A. Exact time of detonation

    B. Location of device

    C. Type of explosive used

    D. All of the above

110. When searching for a possible device, you should:

    A. Keep an eye on the time

    B. Use all your senses

    C. Check for signs of tampering or unusual object placements

    D. All of the above

111. Low explosives have such characteristics as:

    A. High weight

    B. Odd smells

    C. A relatively slow rate of conversion or reaction

    D. None of the above

112. In reacting to an actual explosion, the following must be done:

    A. Evacuate the area, as quickly and safely as possible

    B. Soak the area with water, to prevent secondary explosive devices from activating

    C. Emergency services should be called, once the scene is secured

    D. None of the above

Fire Prevention, Detection, and Response

113. One component of the fire triangle is fuel.

    A. True   **B.** False

114. Class *B* fires are those involving electricity.

   A. True   **B.** False

115. Cluttered areas often have a higher probability of fire hazards.

   A. True   **B.** False

116. Any extinguisher may be safely used on a Class *C* fire.

   A. True   **B.** False

117. Fire plans should:

   A. Be reviewed and supported by management

   B. Be realistic

   C. Be approved by the Fire Marshal

   D. All of the above

118. The fire triangle:

   A. Denotes the area where fires occur

   B. Is a representation of what is required for a fire to exist

   C. Is a chart of which extinguishers apply to a given fire type

   D. None of the above

119. Electronic-detection equipment can check for:

   A. Smoke

   B. Rapid increases in temperature

   C. High temperatures

   D. All of the above

120. Examples of permanent extinguishing hardware might include:

   A. The fire department

   B. Fire extinguishers

   C. Sprinkler systems

   D. None of the above

Hazardous Materials

121. The majority of chemicals and other substances considered hazardous materials:

   A. Are controlled by national laws

   B. Must be transported improperly

   C. Are not inherently dangerous in their original state

   D. Are designed to be out of the reach of children

122. Ultimately, all uncontrolled releases can be traced to:

   A. Equipment failure

   B. Human failure

   C. Improperly followed safety procedures

   D. Lack of hazardous material facilities

123. For decades, the common method of response to a hazardous material release was to:

   A. Notify the local fire department or plant fire brigade

   B. Call the local police department

   C. Wash the contaminated area

   D. Get as much citizen involvement as possible

124. The term "Site Security" refers to:

   A. Sealing off the area, pending an investigation of the incident

   B. Keeping onlookers and bystanders out of the contaminated area

   C. Designating simple entry and exit points

   D. Security that prevents spills

125. The highest area of contamination is called:

   A. The hot zone

   B. The contamination reduction zone

   C. The exclusion zone

   D. The critical zone

126. Nonessential personnel may be allowed at the command post.

   A. True   **B.** False

127. The entire clean-up process must never take more than eight hours.

    A. True   **B.** False

128. The first thing that should be done in an uncontrolled hazardous material release is to notify site personnel about the release.

    A. True   **B.** False

Protection Officers and Emergency Response: Legal and Operational Considerations

129. AED stands for which of the following?

    A. Automated external defibrillator

    B. Altered essential defibrillator

    C. Automated external dysfunction

    D. Automated exterior defibrillator

130. The key to quick emergency medical response is CHECK, CALL, CARE.

    A. True   **B.** False

131. The key to fire safety is the RACE method.

    A. True   **B.** False

132. RACE implies which of the following?

    A. Rescue, Alarm, Confine, Extinguish

    B. Recover, Assist, Constrict, Equip

    C. Rest, Alarm, Confine, Extinguish

    D. Respond, Assist, Constrict, Extinguish

133. The key to hazardous or biohazardous materials (HAZMAT) incidents is to contact the proper agencies and contain the affected area.

    A. True   **B.** False

134. EMS refers to which of the following?

    A. Emergency Medical Student

    B. Exterior Medical Surroundings

    C. Emergency Medical Services

    D. Extensive Mental Services

135. Prior to calling 911 or your local number, the officer should begin to assist the victim with treatment.

    A. True   **B.** False

136. Protection Officers should have first-aid training because they might be able to save lives if an emergency arises at their place of employment.

    A. True   **B.** False

Unit Six

Strikes, Lockouts, and Labor Relations

137. One of the primary functions of a Protection Officer during a strike is picket line surveillance.

    A. True   **B.** False

138. Normally, the Senior Site Security Supervisor is responsible for all security shift responsibilities.

    A. True   **B.** False

139. The key to good security in labor disputes is apprehension, not prevention.

    A. True   **B.** False

140. An effective company-search program may decrease employee morale.

    A. True   **B.** False

141. A suspension provides the employee an opportunity to think about the infraction(s) committed and whether they want to continue employment with the company.

    A. True   **B.** False

142. An effective company-search program can help a company protect its assets by:

    A. Reducing accident rates

    B. Reducing theft

    C. Reducing the use or possession of contraband on property

    D. All of the above

    E. None of the above

143. Which of the following is not considered a type of discipline in labor relations?

    A. A written warning

    B. A suspension

    C. A layoff

    D. A demotion

    E. A discharge

144. Documentation of illegal activities on the picket line could be useful in the following instances:

    A. To support criminal charges

    B. To support company discipline imposed on an employee

    C. To support or defend against unfair labor-practice complaints

    D. To support obtaining an injunction

    E. All of the above

Workplace Violence

145. It was reported that homicide was the fifth-largest cause of occupational-injury death in the workplace.

    A. True   **B.** False

146. Employees working in the retail-trade sector account for what percentage of workplace homicides?

    A. 10 percent

    B. 50 percent

    C. 33 percent

    D. 66 percent

147. According to the United States Office of Personnel Management's Office of Workforce Relations, employee training is _____ component of any prevention strategy.

    A. A critical

    B. A very important

    C. A low priority

    D. An important

148. Most potential workplace violence incidents are preventable if:

    A. Workers who display warning signs of workplace violence are immediately fired.

    B. Proper intervention is achieved early in the escalation process.

    C. Workers who display personal problems are ignored because you never know what will set them off.

    D. Workers who display warning signs of workplace violence are transferred to another area of the company.

149. A workplace violence incident can be perpetrated by either an internal or an external source.

    A. True   **B.** False

150. Only a very small percentage of employees who have violent propensities actually perform a violent crime.

    A. True   **B.** False

151. Workplace violence has only just started to occur in organizations during the last decade.

    A. True   **B.** False

152. Most workplace violence incidents are committed by individuals who are external to the organization (nonemployees).

    A. True   **B.** False

Employee Dishonesty and Crime in Business

153. Which of the following employees steal?

    A. Managers

    B. Supervisors

    C. Line employees

    D. B and C only

    E. A, B, and C

154. Waste containers are the favorite stash places for employees who steal.

   A. True   **B.** False

155. The first step in employee-theft prevention is to learn what can be stolen.

   A. True   **B.** False

156. Protection Officers who miss an assigned round are examples of which WAECUP threat?

   A. Waste

   B. Error

   C. Crime

   D. Unethical/unprofessional practices

   E. Accident

157. It is better to catch employee thieves than to reduce the opportunity for theft.

   A. True   **B.** False

158. It is important to know who is authorized to take trash outside.

   A. True   **B.** False

159. Employee thieves remove company property:

   A. In their own vehicles

   B. In company vehicles

   C. By walking out with it

   D. All of the above

160. First observe, and then report.

   A. True   **B.** False

Substance Abuse

161. Substance abusers only steal from their employer.

   A. True   **B.** False

162. "Crack" is the smoked form of:

    A. Marijuana

    B. Methamphetamine

    C. Valium

    D. Cocaine

163. Marijuana users smoke "sustained low dosages" while at work to avoid detection.

    A. True   B. False

164. Which is an investigative technique used to detect drug dealing on the job?

    A. Undercover investigator

    B. Hidden cameras

    C. Employee interviews

    D. All of the above

165. Workplace drug dealers generally sell their drugs in bathrooms, parking lots, vehicles, and secluded areas.

    A. True   B. False

166. LSD and heroin fall under the same drug category.

    A. True   B. False

167. Drug-abuse prevention is the total responsibility of management.

    A. True   B. False

168. One of the key individuals in a counter-drug-abuse program is the Line Supervisor.

    A. True   B. False

Unit Seven

Effective Communications

169. "Faithful reproduction" of a message means:

    A. A spiritual expression of concern that is understood by all who receives it

    B. A good copy of a message distributed in a timely fashion

    C. A message received and understood by a targeted audience, which contains the exact content of the original message

    D. None of the above

170. To prevent tort-law expectations, which of the following practices should a company pursue?

    A. Conduct perpetual risk analysis

    B. Focus on foreseeability

    C. Mitigate known threats

    D. All of the above

171. Which of the following are the three formal and official forms of communication channels?

    A. Top-down, grapevine, bottom-up

    B. Bottom-up, horizontal, top-down

    C. Grapevine, horizontal, top-down

    D. Top-down, bottom-up, horizontal

172. Effective communications includes all of the following essentials, except:

    A. Content that is factual

    B. Content that expresses beliefs and feelings

    C. Content that is clear and concise

    D. None of the above

173. If you want to achieve clarity in your communications, which of the following will you have to do?

    A. Speak slowly

    B. Choose your words carefully

    C. Select words that are commonly known

    D. All of the above

174. When a receiver hears and understands a message over a two-way radio, they are required to inform the communicator that they heard and understood the message. Which of the following is a correct way to accomplish that confirmation?

    A. Roger

    B. Good Copy

    C. 10–4

    D. All of the above

175. The brevity of the message is important in many situations, such as:

    A. When using duress codes

    B. When using two-way radios

    C. When confirming that you received a message

    D. All of the above

176. Auxiliary equipment commonly found on the security office general use telephone includes which of the following?

    A. A ringer and strobe

    B. A caller ID

    C. A tape-recording device

    D. All of the above

Crisis Intervention

177. Crisis-intervention techniques are designed to provide more control of the eventual outcome of a crisis incident.

    A. True   B. False

178. *Proxemics* refers to how we deliver our words or verbally intervene.

    A. True   B. False

179. The objective of any crisis-development situation is to defuse the situation, while maintaining the safety and welfare of all involved.

    A. True   B. False

180. Anger and frustration are normal reactions of the Protection Officer after a crisis situation.

    A. True   B. False

181. It is better to handle a crisis situation one-on-one; a group would only increase tension.

    A. True   **B.** False

182. During a crisis-development situation, there are four distinct and identifiable behavior levels. This list does not include:

    A. Anxiety

    B. Defensive

    C. Disinterest

    D. Anger/frustration

183. During the evaluation stage of management of disruptive behavior, the first thing the Protection Officer must do is:

    A. Physically restrain the individual

    B. Clear the area of spectators

    C. Implement an action plan

    D. Find out what is happening

184. Personal space is generally defined as:

    A. 1.5 to 3 feet from the individual

    B. The area dictated by the individual

    C. The room occupied by the individual

    D. An arm's length away

Security Awareness

185. Successes in a program should not be publicized, as they might reveal secrets.

    A. True   **B.** False

186. A formal program should be visible and active.

    A. True   **B.** False

187. Newsletters can be used to congratulate specific individuals.

    A. True   **B.** False

188. Smoking near a "No Smoking" sign is an example of failing to enforce policies.

    A. True   **B.** False

189. Involved employees often contribute:

    A. Expertise the security team might not have

    B. A point-of-view that is unique to their working area

    C. An understanding of fellow employees

    D. All of the above

190. A well-designed program:

    A. Is expensive

    B. Is essential

    C. Is bulky

    D. All of the above

191. An example of a "First Moment of Contact" tactic might be:

    A. Signs on the cafeteria tables

    B. Applications that state criminal-background checks are performed

    C. Checking ID tags in the hallways of high-security areas of buildings

    D. None of the above

192. An example of a "Continuation of Contact" tactic might be:

    A. Orientation materials for all new hires

    B. Tall fences at the property line

    C. Seminars

    D. All of the above

Environmental Crime Control Theory

193. There are _____ components of Situational Crime Prevention, Part II.

    A. Two

    B. Four

    C. Eight

    D. Sixteen

194. Further, the components have _____ different subcategories.

   A. Two

   B. Four

   C. Six

   D. Eight

195. Defensible Space: Crime Prevention Through Urban Design was created by:

   A. Oscar Newman

   B. Marcus Felson

   C. Patricia Brantingham

   D. Ronald V. Clarke

196. This revolves around public housing and seeks to reduce crime through the use of natural surveillance, natural access control, and territorial concern.

   A. Crime Prevention Through Environmental Design (CPTED)

   B. Crime Pattern Theory

   C. Defensible Space

   D. Rational Choice Theory

197. Routine Activity Theory, developed by Cohen and Felson, revolves around three things. Which of the following is not one of the three factors?

   A. A potential offender

   B. A suitable target

   C. Acting under the influence of drugs or alcohol

   D. The absence of a capable guardian

198. Routine Activity Theory, developed by Felson and Cohen, revolves around three things: those being a potential offender, a suitable victim or target, and the absence of a capable guardian.

   A. True   **B.** False

199. Developed by _____, Crime Pattern Theory is a complex amalgamation of rational choice, routine activity, and a further introduction of socio-cultural, economic, legal, and the physical environmental cues.

   A. Ronald V. Clarke and Derek B. Cornish

   B. Paul and Patricia Brantingham

   C. Oscar Newman

   D. Ray Jeffery

200. Regarding diffusion of benefits, just as it is assumed by critics of Rationalism that crime is simply moved to another location, there is also a belief that the benefits of situational crime prevention techniques are also moved to other locations, thereby resulting in reduction in crime.

   A. True   **B.** False

Unit Eight

Operational Risk Management

201. Reducing loss and the probability of loss in the organization are whose responsibilities?

   A. The organization's Chief Security Officer

   B. The organization's President or Chief Executive Officer

   C. All personnel in the Security Department

   D. Everyone in the organization

202. An organizational Operational Risk Management program is good to have in place, but it is not really an essential part of the risk identification and mitigation process.

   A. True   **B.** False

203. ORM refers to:

   A. Operational Record Management

   B. Operational Restraint Management

   C. Operational Risk Management

   D. Organizational Risk Maneuvers

204. ORM is vital to the protection of people, property, and information.

   A. True   **B.** False

205. If ORM is to be successful, it must be fully integrated into the organization's culture as a standard way of doing business.

    A. True   **B.** False

206. The Risk Assessment Code (RAC) is used to define the degree of risk associated with a risk and considers incident severity and incident probability. The RAC is derived by using the RAM. How many levels of risk does this include?

    A. Three

    B. Five

    C. Two

    D. Seven

207. Once initial training of all affected personnel is completed on the topic of ORM, it is unnecessary to conduct any periodic training.

    A. True   **B.** False

208. The ORM process exists on how many levels?

    A. Five

    B. Seven

    C. Three

    D. Ten

Emergency Planning and Disaster Control

209. In the event of a disaster, the following authorities should become involved as quickly as is practically possible. (Select the out-of-place item.)

    A. Fire department

    B. Media department

    C. Red Cross

    D. Police department

210. To ensure effective implementation of a disaster plan, it is important that: (Select the best answer.)

    A. Each department head has authority to activate the plan

    B. One individual will be responsible

    C. The disaster team is on call at all times

    D. The Chief Executive Officer or that individual's assistant is available

211. It is important to exclude government authorities when developing a local facility disaster-recovery plan.

   A. True   **B.** False

212. The Disaster Advisory Committee should include key personnel from the fire, safety, and security departments, as well as other departments' personnel.

   A. True   **B.** False

213. An emergency plan and a disaster control program must be flexible enough to meet a variety of complex emergency situations, either those that are manmade or acts of God. Which emergency situation is not considered an act of God?

   A. Earthquake

   B. Flood

   C. Terrorist Act

   D. Tornado

214. In the event of a disaster, it may be necessary to shut down or limit some facility activities because of: (Select the out-of-place item.)

   A. The extent of damage to the facility

   B. The availability of a workforce

   C. The extent and effect of adverse media reports

   D. The availability of internal and external protection units

215. The communications of a warning or alarm must be capable of transmitting throughout the entire facility.

   A. True   **B.** False

216. Security personnel should have access to a current list (including a list of the residents' telephone numbers) of key individuals and organizations that would be involved in the activation of a disaster plan. This list should include:

   A. Corporation department heads

   B. All employees

   C. Police and fire departments

   D. A and C above

Terrorism

217. There are two types of terrorist groups: left-wing terrorist groups and right-wing terrorist groups.

    A. True   **B.** False

218. On what date did the United States encounter the first incident of modern-day left-wing terrorism?

    A. 1953

    B. 1973

    C. 1905

    D. 1961

219. Since 1961, approximately how many major terrorism events have occurred around the world?

    A. 60

    B. 50

    C. 150

    D. 300

220. IRA refers to:

    A. Internal Retaliation Agency

    B. Irish Republican Army

    C. Israel Republican Army

    D. Internal Republican Agency

221. An example of a Hate Group might be the KKK.

    A. True   **B.** False

222. The Al-Qaeda Group would fall under what category?

    A. Hate Group

    B. Left-wing

    C. Right-wing

    D. Militia

223. Right-wing terrorist groups are also known as domestic terrorist groups.

    A. True   **B.** False

224. Eco-terrorist or extremist animal-rights groups are part of the left-wing terrorist group.

    A. True   **B.** False

Counter Terrorism and VIP Protection

225. Terrorists can be _____ by the use of barriers, locks, and response forces.

    A. Denied

    B. Detected

    C. Deterred

    D. Delayed

226. EAS stands for _____.

    A. Electronic Article Surveillance

    B. Entry Article Surveys

    C. Electrical Art Surveys

    D. Expanded Article Surveillance

227. When discussing a search with a searchee, your conversation should be:

    A. Tough and discreet

    B. Polite, considerate, and courteous

    C. Hard and bold

    D. Unprofessional

228. A Personal Protection Specialist should be _____.

    A. A trained teacher first and a trained killer second

    B. A trained killer first and a trained specialist second

    C. A security practitioner first and a trained killer second

    D. A police officer first and a military practitioner second

229. _____, _____, and decorum will "make or break" a PPS quicker than anything else.

    A. Money, fame

    B. Good looks, age

    C. Manners, deportment

    D. IQ, manners

230. According to the U.S. Nuclear Regulatory Commission, one of the following procedures should not be taken:

    A. Stay calm.

    B. Record precise details of the call.

    C. Notify the central alarm station or the Security Shift Supervisor.

    D. Negotiate with the terrorist to maintain calmness.

231. What is a correct procedure if a Security Officer is taken as a hostage or is in close proximity of a hostage taker?

    A. Make a bargain with the hostage taker.

    B. Do not do anything to excite or aggravate the hostage taker.

    C. Do not identify yourself.

    D. Do not attempt to analyze the situation.

232. Communications with the hostage taker should be set up _____.

    A. Immediately

    B. Within 48 hours

    C. Whenever the hostage taker communicates

    D. After 36 hours

Weapons of Mass Destruction: The NBC Threats

233. *Pathogens* are substances produced by living organisms.

    A. True   **B.** False

234. A "dirty bomb" is one that utilizes substandard nuclear material to cause the detonation.

    A. True   **B.** False

235. Nonpersistent agents usually pose a hazard only when the potential for inhalation exists.

    A. True   **B.** False

236. Good security screening could detect WMD threats, although the devices are often difficult to identify.

    A. True   **B.** False

237. Biological weapons often do not require a vast amount of material to impact a wide area or number of people.

    A. True   **B.** False

238. *Ebola* is a form of:

    A. Chemical

    B. Toxin

    C. Pathogen

    D. Fallout

239. Medical treatments may assist in what types of WMD incidents?

    A. Nuclear

    B. Biological

    C. Chemical

    D. All of the above

240. The initial point of contamination may show what signs or impacts?

    A. Lack of animal life

    B. Damage to plant life

    C. Neither A nor B

    D. Both A and B

Unit Nine

Crime Scene Procedures

241. It is important for the Protection Officer to summon enough assistance to:

    A. Properly protect the crime scene

    B. Offer advice on correct procedures

    C. Help gather vital evidence

    D. All of the above

    E. None of the above

242. The reason the Protection Officer must quickly attend all crime scenes is:

    A. To preserve all possible evidence

    B. To show the client they are efficient

    C. To quickly give chase to the culprit

    D. All of the above

243. The moment an officer arrives at a crime scene, they should consider:

    A. Precautions to ensure personal safety

    B. Injured persons

    C. Notes/information

    D. All of the above

    E. None of the above

244. Items that should be recorded in the officer's notebook at a crime scene include:

    A. The date and time of the officer's arrival

    B. Persons present

    C. Date and time of the occurrence

    D. All of the above

    E. None of the above

245. Common forms of physical evidence that could be encountered include: (Select the best answer.)

    A. Tool impressions

    B. Cut wires

    C. Dirt and soil

    D. All of the above

    E. None of the above

246. The main purpose of collecting evidence is to: (Select the best answer.)

    A. Enter it as evidence

    B. Aid in the identification and conviction of the accused

    C. Ensure that curious persons do not contaminate it

    D. All of the above

    E. None of the above

247. Once the boundaries of a crime scene are established, it is less important to keep people away from the area.

    A. True   B. False

248. Fellow officers and occupants at the crime scene need not be excluded from the scene once the boundaries are established.

    A. True   B. False

Foundations for Surveillance

249. Overt operations are not designed to be actively hidden.

    A. True   B. False

250. Personal operations allow for human interpretation of events and active decision-making.

    A. True   B. False

251. Visual records should only be viewed once to preserve clarity of thought.

    A. True   B. False

252. Privacy issues are best left to the company lawyer in reviewing the case after the evidence is compiled.

    A. True   **B.** False

253. A uniformed patrol officer is an example of:

    A. Overt personal operations

    B. Covert personal operations

    C. Overt electronic operations

    D. Covert electronic operations

254. A man wearing street clothes and in an unmarked car is an example of:

    A. Overt personal operations

    B. Covert personal operations

    C. Overt electronic operations

    D. Covert electronic operations

255. A camera built into the background or in a clock face is an example of:

    A. Overt personal operations

    B. Covert personal operations

    C. Overt electronic operations

    D. Covert electronic operations

256. A dome camera, mounted next to the exit door of a warehouse, is an example of:

    A. Overt personal operations

    B. Covert personal operations

    C. Overt electronic operations

    D. Covert electronic operations

Interviewing Techniques

257. If the Protection Officer is unable to control the interview:

    A. Time will be lost

    B. Facts may be forgotten

    C. Psychological advantage shifts

    D. All of the above

258. An experienced interviewer need not have a game plan when interviewing.

    A. True   **B.** False

259. The success or failure of an interview depends entirely on the skill of the investigator.

    A. True   **B.** False

260. The location of the interview should be chosen by the subject, so they will be more at ease in familiar surroundings.

    A. True   **B.** False

261. In most instances, the content of an incident will be covered in more than one conversation.

    A. True   **B.** False

262. Rapid-fire questioning techniques will most likely:

    A. Cause the subject to tell the truth

    B. Confuse the subject

    C. Cause the subject to lie

    D. B and C above

263. By asking a series of questions early in the interview, you:

    A. Condition the subject to believe that, if you want information, you will ask

    B. Lead the subject to believe that everything they tell you has significance

    C. Cause the subject to withhold information by putting them on guard

    D. All of the above

264. Obstacles to conversations in an interview are:

    A. Specific questions

    B. Yes and no questions

    C. Use of leading questions

    D. All of the above

Investigation: Concepts and Practices for Security Professionals

264. Investigation comes from the Latin word _____.

    A. investigere

    B. anvestigere

    C. devestigere

    D. investgote

266. The contemporary Protection Officer acts as a Management Representative, a Legal Consultant, an Enforcement Agent, and a _____.

    A. Store Manager

    B. Retail Loss Supervisor

    C. Homicide Detective

    D. Intelligence Agent

267. Investigation and _____ go hand-in-hand.

    A. Asset protection

    B. Retail loss

    C. Guidelines

    D. Inductive reasoning

268. Investigators use two types of logic: inductive reasoning and _____.

    A. Investigation reasoning

    B. Detective reasoning

    C. Deductive reasoning

    D. Logical reasoning

269. Who originated the saying: "It takes a thief to catch a thief"?

    A. Allan Pinkerton

    B. Jonathon Wild

    C. Charles Dickens

    D. Howard Vincent

270. Who was the attorney placed in charge of the Scotland Yard in 1878?

    A. Charles Dickens

    B. John Edgar Hoover

    C. Allan Pinkerton

    D. Howard Vincent

271. What decision required all law-enforcement personnel in the U.S. to advise suspects of their rights before asking them any questions?

    A. The Rights decision

    B. The Miranda decision

    C. The Morales decision

    D. The Exclusionary Rule decision

272. There are some things to bear in mind when you testify in court. One that should *not* be taken is:

    A. Always be positive

    B. Do not project your voice to the jury

    C. Be neat, clean, and conservatively dressed

    D. Have the case prepared for trial

Unit Ten

Legal Aspects of Security

273. The purpose of our legal system is to:

    A. Set down our obligations to each other

    B. Set penalties for breaching these obligations

    C. Establish procedures to enforce these obligations

    D. All of the above

274. The common law never changes.

    A. True   **B.** False

275. The doctrine of case law states that a court must stand by previous decisions.

    A. True   **B.** False

276. Statutes are changed:

   A. Never

   B. To fill a need in our society

   C. Only by a level of government higher than the one that passed the law

   D. Whenever there is a change in government

277. The prosecutor's job is to get compensation for the victim.

   A. True   **B.** False

278. The police will investigate:

   A. Civil matters

   B. Criminal matters

   C. Whatever they are paid to investigate

   D. All of the above

279. A civil action may not commence until the criminal courts are finished.

   A. True   **B.** False

280. A warrant to arrest may be executed by:

   A. A private citizen

   B. The police

   C. A Security Officer

   D. Anyone who apprehends the suspect

Unit Eleven

Use of Force

281. A subject may admit wrongdoing during what period?

   A. Debriefing

   B. Physical confrontation

   C. Assault

   D. Reprimand

282. Calmness is:

   A. Contrasting

   B. Contagious

   C. Lethal

   D. Tactical

283. The definition of "reasonable use of force" is the amount of force:

   A. Necessary to hurt the aggressor

   B. Needed to incapacitate the aggressor

   C. Equal to or slightly greater than the force of the aggressor

   D. Necessary to seriously scare the aggressor

284. When writing a report in use of force cases, it is appropriate to do all the following, except:

   A. The report follows a chronology

   B. The report has contradictions

   C. Times stated match other reports and records

   D. Facts stated match other reports and records

285. Which of the following is a recommended formula for self-control?

   A. Intellect/Emotions = Control

   B. Emotions/Intellect = Control

   C. Training/Intellect = Control

   D. Problem/Emotions = Control

286. "Tachy-psyche effect" can be identified from:

   A. Proper background investigation

   B. Physiological characteristics

   C. Family background

   D. Ethnic or racial make-up

287. When approaching apparently rational subjects, officers should do all of the following, except:

    A. Respect the subjects' dignity

    B. Become part of the problem

    C. Shout commands

    D. Be aggressive

288. What is the cornerstone of officer survival?

    A. The use of force

    B. Crisis intervention

    C. Conflict resolution

    D. Aggressive behavior

Defensive Tactics and Officer Safety

289. The basic goal of defensive tactics is to avoid injury to yourself and others.

    A. True   **B.** False

290. Using all your senses helps to maintain awareness of your surroundings.

    A. True   **B.** False

291. Control of timing is a reference to knowing when to patrol certain areas during certain times of the day.

    A. True   **B.** False

292. The use of "Weapon-Assisted Controls" is generally considered a higher level of force than Striking Techniques.

    A. True   **B.** False

293. Staying to the side of your opponent is an example of:

    A. Control of timing

    B. Control of space

    C. Control of damage

    D. None of the above

294. Choosing to disable or strike your opponent in the leg to allow you to escape is an example of:

    A. Control of timing

    B. Control of space

    C. Control of damage

    D. None of the above

295. Selecting a linear strike, as opposed to a wide and sweeping strike, is an example of:

    A. Control of timing

    B. Control of space

    C. Control of damage

    D. None of the above

296. In the typical physical confrontation, which concept is not utilized?

    A. Control of timing

    B. Control of space

    C. Control of damage

    D. All are utilized

Apprehension and Detention Procedures

297. *Black's Law Dictionary* defines arrest as depriving a person of their liberty by legal:

    A. Document

    B. Action

    C. Authority

    D. Declaration

298. All that is required for an arrest is some act by a Security Officer indicating their intention to detain.

    A. True   **B.** False

299. A security company may have a contractual agreement for an officer to exercise a certain level of authority from the:

    A. Government

    B. Media

    C. Public

    D. Client

300. As with most things legal, the word "reasonable" keeps popping up in discussions about apprehension and detention.

    A. True   **B.** False

301. Whenever physical contact is made with a citizen, the Security Officer should:

    A. File a complaint

    B. Write a detailed report

    C. Contact an attorney

    D. File a grievance

302. A violation of the criminal law is known as a tort.

    A. True   **B.** False

303. A crucial part of avoiding criminal and civil liability is:

    A. Good human-relations skills

    B. A good alibi

    C. Good communication skills

    D. A and C

304. The Latin term "non solis" means "never alone."

    A. True   **B.** False

Unit Twelve

Public Relations

305. Security Officers develop the "malcontent syndrome" by working shorter hours, having days off, and working on good assignments.

    A. True   **B.** False

306. To be prepared for the unexpected, Security Officers should always carry a flashlight, a watch, a notepad, and:

    A. A screwdriver

    B. A pocket knife

    C. A two-way radio, ready in hand

    D. Identification, hanging on their neck

307. To show interest when greeting a customer, a great opening statement would be:

    A. May I help you?

    B. I'll be there in one sec, hang on.

    C. Hi! What can I get for you?

    D. How may I help you?

308. The successful Protection Officer will know every other department's products and services.

    A. True   **B.** False

309. Which of the following selections is an in-depth evaluation of identified threats, probability hypotheses, vulnerability studies, and security surveys of facilities and systems?

    A. Media relations

    B. Supervisor analysis

    C. Training reports

    D. Risk analysis

310. Whenever possible:

    A. Stay out of people's business

    B. Let the supervisor handle the situation

    C. Be polite, but distant

    D. Help others as much as possible

311. Present a professional appearance:

    A. When meeting the press

    B. Only during inspections

    C. Only when you have to

    D. At all times

312. Security personnel must be:

    A. Managers

    B. Teachers

    C. Salespeople

    D. Instructors

Police and Security Liaison

313. Relationships between law enforcement and private security in the early 1980s were rated fair to good.

　　A. True　　　　　　　　**B.** False

314. In 1976, the Private Security Advisory Council, through the U.S. Department of Justice, identified two main factors that contributed to poor relationships between law enforcement and private security. (Identify the two correct factors.)

　　A. Their inability to clarify role definitions.

　　B. They often practiced stereotyping

　　C. Lack of funding

　　D. Resentment

315. Liaison plays a less-than-substantial role in our daily functions as security professionals.

　　A. True　　　　　　　　**B.** False

316. Security officers will outnumber police officers 3 to 1.

　　A. True　　　　　　　　**B.** False

317. The cost of economic crime is over:

　　A. $110 million

　　B. $120 million

　　C. $1 billion

　　D. $114 billion

318. According to the Hallcrest Report II, Private Security Trends (1970 to 2000), private security is America's secondary protective resource in terms of spending and employment.

　　A. True　　　　　　　　**B.** False

319. There appears to be a growing potential for contracting private security to perform the following activities: (Select the incorrect activity.)

　　A. Courtroom security

　　B. Special-event security

　　C. Hostage negotiations

　　D. Traffic control

320. The key to maintaining a good working relationship with law-enforcement personnel is to: (Select the best answer.)

    A. Maintain a good public perception

    B. Enroll in law-enforcement programs

    C. Send them all your reports

    D. Maintain a high level of physical fitness

Ethics and Professionalism

321. Bottom and Kostanoski's WAECUP model asserts that losses stem from all of the following, except:

    A. Waste

    B. Accident

    C. Crime

    D. Professional practices

322. Which is not in the International Foundation for Protection Officers' Code of Ethics?

    A. Respond to the employer's professional needs

    B. Respond to the employer's personal needs

    C. Protect confidential information

    D. Dress to create professionalism

323. Which of the following is not true?

    A. The *P* in "professional" is for precise.

    B. The *E* in "professional" is for ethical.

    C. The *I* in "professional" is for informed.

    D. The *A* in "professional" is for accompany.

324. Professionals always think in terms of their:

    A. Looks

    B. Duties

    C. Authority

    D. Image

325. The acronym PORT includes all but:

 A. Personal

 B. Options

 C. Responsibilities

 D. Time

326. The letter *R* in the acronym PORT stands for:

 A. Review

 B. Responsibility

 C. Respond

 D. None of the above

327. Deportment can be defined as:

 A. Making a poor decision

 B. A continuous process

 C. Writing out the PORT

 D. How one carries oneself

328. Manners that promote a professional image do not include:

 A. Standing to greet people

 B. Praising others when appropriate

 C. Allowing people to talk without interruption

 D. Tucking a napkin into your belt

The International Foundation for Protection Officers does not issue or endorse the Registered Security Professional certificate. To receive your Registered Security Professional certificate, submit your answers to: IAHSSP; Box 2044; Erie, PA 16512.

1. _____	36. _____	71. _____
2. _____	37. _____	72. _____
3. _____	38. _____	73. _____
4. _____	39. _____	74. _____
5. _____	40. _____	75. _____
6. _____	41. _____	76. _____
7. _____	42. _____	77. _____
8. _____	43. _____	78. _____
9. _____	44. _____	79. _____
10. _____	45. _____	80. _____
11. _____	46. _____	81. _____
12. _____	47. _____	82. _____
13. _____	48. _____	83. _____
14. _____	49. _____	84. _____
15. _____	50. _____	85. _____
16. _____	51. _____	86. _____
17. _____	52. _____	87. _____
18. _____	53. _____	88. _____
19. _____	54. _____	89. _____
20. _____	55. _____	90. _____
21. _____	56. _____	91. _____
22. _____	57. _____	92. _____
23. _____	58. _____	93. _____
24. _____	59. _____	94. _____
25. _____	60. _____	95. _____
26. _____	61. _____	96. _____
27. _____	62. _____	97. _____
28. _____	63. _____	98. _____
29. _____	64. _____	99. _____
30. _____	65. _____	100. _____
31. _____	66. _____	101. _____
32. _____	67. _____	102. _____
33. _____	68. _____	103. _____
34. _____	69. _____	104. _____
35. _____	70. _____	105. _____

106. _____	146. _____	186. _____
107. _____	147. _____	187. _____
108. _____	148. _____	188. _____
109. _____	149. _____	189. _____
110. _____	150. _____	190. _____
111. _____	151. _____	191. _____
112. _____	152. _____	192. _____
113. _____	153. _____	193. _____
114. _____	154. _____	194. _____
115. _____	155. _____	195. _____
116. _____	156. _____	196. _____
117. _____	157. _____	197. _____
118. _____	158. _____	198. _____
119. _____	159. _____	199. _____
120. _____	160. _____	200. _____
121. _____	161. _____	201. _____
122. _____	162. _____	202. _____
123. _____	163. _____	203. _____
124. _____	164. _____	204. _____
125. _____	165. _____	205. _____
126. _____	166. _____	206. _____
127. _____	167. _____	207. _____
128. _____	168. _____	208. _____
129. _____	169. _____	209. _____
130. _____	170. _____	210. _____
131. _____	171. _____	211. _____
132. _____	172. _____	212. _____
133. _____	173. _____	213. _____
134. _____	174. _____	214. _____
135. _____	175. _____	215. _____
136. _____	176. _____	216. _____
137. _____	177. _____	217. _____
138. _____	178. _____	218. _____
139. _____	179. _____	219. _____
140. _____	180. _____	220. _____
141. _____	181. _____	221. _____
142. _____	182. _____	222. _____
143. _____	183. _____	223. _____
144. _____	184. _____	224. _____
145. _____	185. _____	225. _____

226. _____
227. _____
228. _____
229. _____
230. _____
231. _____
232. _____
233. _____
234. _____
235. _____
236. _____
237. _____
238. _____
239. _____
240. _____
241. _____
242. _____
243. _____
244. _____
245. _____
246. _____
247. _____
248. _____
249. _____
250. _____
251. _____
252. _____
253. _____
254. _____
255. _____
256. _____
257. _____
258. _____
259. _____
260. _____
261. _____
262. _____
263. _____
264. _____
265. _____

266. _____
267. _____
268. _____
269. _____
270. _____
271. _____
272. _____
273. _____
274. _____
275. _____
276. _____
277. _____
278. _____
279. _____
280. _____
281. _____
282. _____
283. _____
284. _____
285. _____
286. _____
287. _____
288. _____
289. _____
290. _____
291. _____
292. _____
293. _____
294. _____
295. _____
296. _____
297. _____
298. _____
299. _____
300. _____
301. _____
302. _____
303. _____
304. _____
305. _____

306. _____
307. _____
308. _____
309. _____
310. _____
311. _____
312. _____
313. _____
314. _____
315. _____
316. _____
317. _____
318. _____
319. _____
320. _____
321. _____
322. _____
323. _____
324. _____
325. _____
326. _____
327. _____
328. _____

Appendix E

REGISTERED PROFESSIONAL LOCKSMITH EXAM

Locksmith and Security Professionals' Exam Study Guide

Exam I

This test is based on the International Association of Home Safety and Security Professionals' Registered Professional Locksmith registration program. If you earn a passing score, you should be able to pass other locksmithing certification and licensing examinations.

1. An otoscope can be helpful for reading disc-tumbler locks by providing light and magnification.

    A. True   **B.** False

2. Kwikset locks come with a KW1 keyway.

    A. True   **B.** False

3. Many Schlage locks come with an SCL1 keyway.

    A. True   **B.** False

4. The purpose of direct (uncoded) codes on locks is to obscure the lock's bitting numbers.

    A. True   **B.** False

5. A skeleton key can be used to open warded bit-key locks.

    A. True   **B.** False

6. Typically, the lock on the front of the car's driver side will be harder to pick open than other less-often-used locks on the car.

    A. True   **B.** False

7. A standard electromagnetic lock includes a rectangular electromagnet and a rectangular wood-and-glass strike plate.

    A. True   **B.** False

8. A *blank* is a key that fits two or more locks.

    A. True   **B.** False

9. One difference between a bit key and barrel key is the barrel key has a hollow shank.

    A. True   **B.** False

10. Parts of a flat key include the bow, blade, and throat cut.

    A. True   **B.** False

11. The Egyptians are credited with inventing the first lock to be based on the locking principle of today's pin-tumbler lock.

    A. True   **B.** False

12. Before impressioning a pin-tumbler cylinder, it's usually helpful to lubricate the pin chambers thoroughly.

    A. True   **B.** False

13. When you are picking a pin-tumbler cylinder, spraying a little lubrication into the keyway may be helpful.

    A. True   **B.** False

14. If a customer refuses to pay you after you finish the job at their house, you have the legal right to remain inside their house until they pay you.

    A. True   **B.** False

15. A long-reach tool and a wedge are commonly used to open locked automobiles.

    A. True   **B.** False

16. It's legal for locksmiths to duplicate a U.S. Post Office box key at the request of the box renter— if the box renter shows a current passport or a driver's license.

    A. True   **B.** False

17. The Romans are credited with inventing the warded lock.

    A. True   **B.** False

18. Five common keyway groove shapes are left angle, right angle, square, *V*, and round.

    A. True   **B.** False

19. To pick open a pin-tumbler cylinder, you usually need a pick and a torque wrench.

    A. True   **B.** False

20. Fire-rated exit devices usually have dogging.

    A. True   **B.** False

21. Common door-lock backsets include:

    A. 2½ × 2½ inches

    B. 2¾ × 2⅜ inches

    C. 1½ × 2½ inches

    D. 2⅜ × 3 inches

22. How many sets of pin tumblers are in a typical pin-tumbler house door lock?

    A. Three or four

    B. Five or six

    C. Eleven or twelve

    D. Seven or eight

23. Which lock is unpickable?

    A. A Medeco biaxial deadbolt

    B. A Grade 2 Titan

    C. The Club steering wheel lock

    D. None of the above

24. Which are basic parts of a standard key-cutting machine?

    A. A pair of vises, a key stop, and a grinding stylus

    B. Two cutter wheels, a pair of vises, and a key shaper

    C. A pair of vises, a key stylus, and a cutter wheel

    D. A pair of styluses, a cutter wheel, and a key shaper

25. What are two critical dimensions for code-cutting cylinder keys?

    A. Spacing and depth

    B. Bow size and blade thickness

    C. Blade width and keyhole radius

    D. Shoulder width and bow size

26. Which manufacturer is best known for its low-cost residential key-in-knob locks?

    A. Kwikset Corporation

    B. Medeco Security Lock

    C. The Key-in-Knob Corporation

    D. ASSA

27. The most popular mechanical lock brands in the United States include:

    A. Yale, Master, Corby, and Gardall

    B. Yale, Kwikset, Master, and TuffLock

    C. Master, Weiser, Kwikset, and Schlage

    D. Master, Corby, Gardall, and TuffLock

28. A mechanical lock operated mainly by a pin-tumbler cylinder is commonly called a:

    A. Disc-tumbler pinned lock

    B. Cylinder pin lock

    C. Mechanical-cylinder pin lock

    D. Pin-tumbler cylinder lock

29. A key-in-knob lock whose default position is that both knobs are locked and require a key to be used for unlocking is:

    A. A classroom lock

    B. A function lock

    C. An institution lock

    D. A school lock

30. Four basic types of keys are:

    A. Barrel, flat, bow, and tumbler

    B. Cylinder, flat, warded, and V-cut

    C. Dimple, angularly bitted, corrugated, and blade

    D. Cylinder, flat, tubular, and barrel

31.   The two most common key stops are:

    A. Blade and V-cut

    B. Shoulder and tip

    C. Bow and blade

    D. Keyway grooves and bittings

32.   Bit keys most commonly are made of:

    A. Brass, copper, and silver

    B. Aluminum, iron, and silver

    C. Iron, brass, and aluminum

    D. Copper, silver, and aluminum

33.   Which of the following key combinations provides the most security?

    A. 55555

    B. 33333

    C. 243535

    D. 35353

34.   Which of the following direct key combinations provides the least security?

    A. 243535

    B. 1111

    C. 321231

    D. 22224

35.   A blank is basically just:

    A. A change key with cuts on one side only

    B. An uncut or uncombinated key

    C. Any key with no words or numbers on the bow

    D. A master key with no words or numbers on the bow

36. You often can determine the number of pin stacks or tumblers in a cylinder by:

    A. Its key blade length

    B. Its key blade thickness

    C. The key blank manufacturer's name on the bow

    D. The material of the key

37. Spool and mushroom pins:

    A. Make keys easier to duplicate

    B. Can hinder normal picking attempts

    C. Make a lock easier to pick

    D. Make keys harder to duplicate

38. As a general rule, General Motors' 10-cut wafer sidebar locks have:

    A. A sum total of cut depths that must equal an even number

    B. Up to four of the same depth cut in the 7, 8, 9, and 10 spaces

    C. A maximum of five number 1 depths in a code combination

    D. At least one 4–1 or 1–4 adjacent cuts

39. When drilling open a standard pin-tumbler cylinder, position the drill bit:

    A. At the first letter of the cylinder

    B. At the shear line in alignment with the top and bottom pins

    C. Directly below the bottom pins

    D. Directly above the top pins

40. When viewed from the exterior side, a door that opens inward and has hinges on the right side is a:

    A. Left-hand door

    B. Right-hand door

    C. Left-hand reverse bevel door

    D. Right-hand reverse bevel door

41. A utility patent:

    A. Relates to a product's appearance, is granted for 14 years, and is renewable

    B. Relates to a product's function, is granted for 17 years, and is nonrenewable

    C. Relates to a product's appearance, is granted for 17 years, and is renewable

    D. Relates to a product's function, is granted for 35 years, and is nonrenewable

42. To earn a UL-437 rating, a sample lock must:

    A. Pass a performance test

    B. Use a patented key

    C. Use hardened-steel mounting screws and mushroom and spool pins

    D. Pass an attack test using common hand and electric tools, such as drills, saw blades, puller mechanisms, and picking tools

43. Tumblers are:

    A. Small metal objects that protrude from a lock's cam to operate the bolt

    B. Fixed projections on a lock's case

    C. Small pins, usually made of metal, that move within a lock's case to prevent unauthorized keys from entering the keyhole

    D. Small objects, usually made of metal, that move within a lock cylinder in ways that obstruct a lock's operation until an authorized key or combination moves them into alignment

44. Electric switch locks:

    A. Are mechanical locks that have been modified to operate with battery power

    B. Complete and break an electric current when an authorized key is inserted and turned

    C. Are installed in metal doors to give electric shocks to intruders

    D. Are mechanical locks that have been modified to operate with alternating-current (AC) electricity instead of with a key

45. A popular type of lock used on GM cars is:

    A. A Medeco pin tumbler

    B. An automotive bit key

    C. A sidebar wafer

    D. An automotive tubular key

46. When cutting a lever-tumbler key by hand, the first cut should be the:

    A. Lever cut

    B. Stop cut

    C. Throat cut

    D. Tip cut

47. How many possible key changes does a typical disk-tumbler lock have?

    A. 1500

    B. 125

    C. A trillion

    D. 25

48. Which manufacturer is best known for its interchangeable core locks?

    A. Best Lock

    B. Kwikset Corporation

    C. Ilco Interchangeable Core Corporation

    D. Interchangeable Core Corporation

49. James Sargent is famous for:

    A. Inventing the Sargent key-in-knob lock

    B. Inventing the time lock for banks

    C. Inventing the double-acting, lever-tumbler lock

    D. Being the first person to pick open a Medeco biaxial cylinder

50. Which are the common parts of a combination padlock?

    A. Shackle, case, bolt

    B. Spacer washer, top pins, cylinder housing

    C. Back cover plate, case, bottom pins

    D. Wheel pack base plate, wheel pack spring, top and bottom pins

51. General Motors' ignition lock codes generally can be found:

 A. On the ignition lock

 B. On the passenger-side door

 C. Below the Vehicle Identification Number (VIN) on the vehicle's engine

 D. Under the vehicle's brake pedal

52. Which code series is commonly used on Chrysler door and ignition locks?

 A. EP 1–3000

 B. CHR 1–5000

 C. CRY 1–4000

 D. GM 001–6000

53. How many styles of lock pawls does General Motors use in its various car lines?

 A. One

 B. Five

 C. Over 20

 D. Three

54. The double-sided (or 10-cut) Ford key:

 A. Has five cuts on each side. One side operates the trunk and door, and the other side operates the ignition.

 B. Has five cuts on each side. Either side can operate all the locks of a car.

 C. Has ten cuts on each side. One side operates the trunk and door, and the other side operates only the ignition.

 D. Has ten cuts on each side.

55. Usually the simplest way to change the combination of a double-bitted cam lock is to:

 A. Rearrange the positions of two or more tumblers

 B. Remove two tumblers and replace them with new tumblers

 C. Remove the tumbler assembly and replace it with a new one

 D. Connect a new tumbler assembly to the existing one

56. When shimming a cylinder open:

    A. Use the key to insert the shim into the keyway.

    B. Insert the shim into the keyway without the key.

    C. Insert the shim along the left side of the cylinder housing.

    D. Insert the shim between the plug and cylinder housing between the top and bottom pins.

57. A lock is any:

    A. Barrier or closure that restricts entry

    B. Fastening device that allows a person to open and close a door, window, cabinet, drawer, or gate

    C. Device that incorporates a bolt, cam, shackle, or switch to secure an object—such as a door, drawer, or machine—to a closed, locked, on, or off position, and that provides a restricted means—such as a key or combination—of releasing the object from that position

    D. Device or object that restricts entry to a given premise

58. Which wheel in a safe lock is closest to the dial?

    A. Wheel 1

    B. Wheel 2

    C. Wheels 3

    D. Wheel 0

59. Which is *not* a type of safe combination wheel?

    A. Hole change

    B. Dial change

    C. Key change

    D. Screw change

60. Which type of cylinder is typically found on an interlocking deadbolt (or "jimmy-proof deadlock")?

    A. Mortise cylinder

    B. Key-in-knob cylinder

    C. Rim cylinder

    D. Tubular deadbolt cylinder

Exam 2

This test is based on the International Association of Home Safety and Security Professionals' Registered Professional Locksmith registration program. If you earn a passing score, you should be able to pass other locksmithing certification and licensing examinations.

1. As a general rule, General Motor's ten-cut wafer side-bar locks:

    A. Have no 4 depth in the first space (closest to the shoulder)

    B. Have at least one 4–1 or 1–4 adjacent cuts

    C. Have a maximum of five number-one depths in a code combination

    D. Have up to four of the same depth cut in the 7, 8, 9, and 10 spaces

2. When drilling a cylinder open:

    A. Position the drill bit at least 1 inch above the cylinder and drill upward

    B. Position the drill bit at the shear line

    C. Position the drill bit at least 1 inch below the cylinder and drill downward

    D. Position the drill at the first letter on the face of the cylinder

3. Chicago Lock Company:

    A. Made the first tubular key lock

    B. Made the first lever tumbler lock

    C. Makes the Chicago "Stamper-bolt" lock

    D. Made the first key-in-knob lock

4. When viewed from the exterior side, a door that opens inward and has hinges on the right is:

    A. A right-hand door

    B. A left-hand door

    C. A left-hand reverse bevel door

    D. A right-hand reverse bevel door

5. When shimming a cylinder open:

    A. Insert the shim into the keyway

    B. Insert the shim along the bottom of the keyway

    C. Insert the shim along the left side of the plug

    D. Insert the shim between the plug and cylinder housing at the pin tumblers

6. To earn the UL burglary safe rating "TRTL-30," a safe:

    A. Must weigh at least 1000 pounds

    B. Must have a door that opens inward

    C. Must have relockers along its floor

    D. Must weigh at least 750 pounds or be equipped with anchors

7. A utility patent:

    A. Relates to a product's appearance, is granted for 14 years, and is renewable

    B. Relates to a product's function, is granted for 17 years, and is nonrenewable

    C. Relates to a product's appearance, is granted for 17 years, and is renewable

    D. Relates to a product's function, is granted for 35 years, and is nonrenewable

8. To earn a UL-437 rating, a lock must:

    A. Pass an attack test using common hand and electric tools, drills, saw blades, puller mechanisms, and picking tools

    B. Pass a performance test only

    C. Use a patented key

    D. Use hardened-steel mounting screws

9. The pin-tumbler lock patented by Linus Yale Jr., in 1865, used a mortise-style cylinder that operated with a flat key.

    A. True   **B.** False

10. In 1868 Henry R. Towne:

    A. Joined with Linus Yale, Jr., to form the Yale Lock Manufacturing Company

    B. Joined with Linus Yale, Sr., to form the Yale and Towne Manufacturing Company

    C. Joined with Walter Schlage to form the Schlage Lock Company

    D. Joined with Stephen Bucknall to form the Eagle Lock Company

11. Fire-rated exit devices usually have dogging.

    A. True   **B.** False

12. A lock is:

    A. Any barrier or closure

    B. Any fastening device that allows a person to open and close a door, window, cabinet, drawer, or gate

    C. Any device that incorporates a bolt, cam, shackle, or switch to secure an object—such as a door, drawer, or machine—to a closed, locked, on, or off position, and that provides a restricted means of releasing the object from that position

    D. Any device or object that restricts entry to a given premises

13. Tumblers are:

    A. Fixed projections on the case of a lock

    B. Small objects that protrude out of a lock's cam to activate the bolt

    C. Small pins, usually made of metal, that move within a bit-key lock's case to prevent unauthorized keys from entering the keyhole

    D. Small objects, usually made of metal, that move within a lock cylinder in ways that obstruct a lock's operation until an authorized key or combination moves them into alignment

14. It's legal to duplicate a United States Post Office box key only after obtaining written permission from the box renter.

    A. True   B. False

15. Electric switch locks:

    A. Are mechanical locks that have been modified to operate with AC electricity instead of with a key

    B. Are mechanical locks that have been modified to operate with battery power

    C. Are installed in metal doors to give electric shocks to intruders

    D. Complete and break an electric current when an authorized key is inserted and turned.

16. Based on ANSI A156 standards, Grade 3 locks are for heavy-duty commercial applications.

    A. True   B. False

17. Worn pin tumblers are generally easier to pick than new pin tumblers.

    A. True   B. False

18. Disc-tumbler, or wafer, locks have two shear lines that tumblers can protrude past to prevent the rotation of a plug.

    A. True   **B.** False

19. When picking a typical pin-tumbler cylinder:

    A. The bottom pins must be raised to the shear line.

    B. The top pins must be pulled below the shear line.

    C. The tension wrench, or turning tool, should be inserted only after all the pins have been picked into a position that frees the plug to be turned.

    D. The pick should be inserted into the keyway only after the tension wrench, or turning tool, has rotated the plug into position and has been removed from the keyway.

20. Before duplicating a General Motors VATS key:

    A. You have to remove the resistor pellet from the key being copied.

    B. You have to demagnetize the resistor pellet in the key being copied.

    C. You need to measure the resistance of the pellet in the key to determine which blank to use.

    D. Use a ground strap to avoid transferring static electricity to the resistor pellets.

21. When picking a pin-tumbler cylinder, sometimes spraying a little lubrication into the keyway can be helpful.

    A. True   **B.** False

22. Before impressioning a pin-tumbler cylinder, it's usually helpful to thoroughly lubricate the pin chambers.

    A. True   **B.** False

23. The older a pin tumbler lock is, the easier it is to impression.

    A. True   **B.** False

24. The most popular type of lock used on GM cars is a:

    A. Side bar lock

    B. Pin tumbler lock

    C. Tubular key lock

    D. Flat key lock

25. If a cut in a key for a pin-tumbler cylinder isn't deep enough:

   A. The bottom pin for that cut will drop below the shear line.

   B. The top pin for that cut will drop below the shear line.

   C. The bottom pin for the adjacent cut will drop below the shear line.

   D. The bottom pin for that cut will rise above the shear line.

26. The bit of a bit key:

   A. Manipulates the tumblers in a warded bit-key lock

   B. Enters the keyway of a cylinder lock

   C. Enters the keyhole of a bit-key lock

   D. Comes into contact with the tailpiece of a bit-key lock to extend and retract the lock bolt

27. Spool and mushroom pins:

   A. Make a lock easier to pick

   B. Make keys harder to duplicate

   C. Make a lock harder to pick

   D. Make keys easier to duplicate

28. An employer who fails to conduct background investigations before hiring employees can be held liable for the criminal acts of the employees.

   A. True   **B.** False

29. Usually, the simplest way to change the combination of a double-bitted cam lock is to:

   A. Rearrange the positions of two or more tumblers

   B. Remove two tumblers and replace them with new tumblers

   C. Remove the tumbler assembly and replace it with a new one

   D. Connect a new tumbler assembly to the existing one

30. The blade length of a key is most indicative of:

   A. The manufacturer of the lock the key is designed to operate

   B. The number of tumblers in the cylinder the key is designed to operate

   C. The lengths of the master pins in the cylinder the key is designed to operate

   D. The number of master pins in the cylinder the key is designed to operate

31. An uncut key is called a:

    A. Key blank

    B. Change key

    C. Master key

    D. Bit key

32. The "hand of a door" refers to:

    A. The location of the hinges and swing of the door

    B. The type of lock on the door

    C. The door-knob side of a door

    D. The direction the door knob turns to open the door

33. Which type of lock generally provides the least security?

    A. Pin tumbler

    B. Deadbolt

    C. Side bar

    D. Warded

34. Which type of lock would probably provide the most security?

    A. ANSI Grade 3 key-in-knob lock with dual-pin tumbler cylinders, five master pins, and hardened-steel bolt

    B. ANSI Grade 2 key-in-knob with dual UL-437 listed cylinders, five master pins, and hardened-steel bolt

    C. ANSI Grade 2 deadbolt lock with UL-437 listed cylinder, patented key, and hardened-steel bolt

    D. Interlocking deadbolt with spool and mushroom bottom pins, five master pins, and a key stamped "Do Not Duplicate"

35. Among common keys, those with the longest life span are made of:

    A. Aluminum

    B. Brass

    C. Copper-silver alloy

    D. Borax iron-steel alloy compounds

36. Which of the following key combinations provides the least security?

  A. 36242

  B. 24263

  C. 11111

  D. 22241

37. Which of the following key combinations provides the most security?

  A. 11111

  B. 252146

  C. 45565

  D. 31222

38. When cutting a lever-tumbler key by hand, the first cut should be the:

  A. Tip cut

  B. Stop cut

  C. Throat cut

  D. Lever cut

39. Successful lockpicking depends primarily on:

  A. The length of the tension wrench

  B. The material of the cylinder

  C. Knowledge about the lock being picked

  D. The length of the lock bolt

40. James Sargent is famous for:

  A. Being the first person to pick a Yale pin-tumbler cylinder

  B. Inventing the double-acting lever tumbler lock

  C. Inventing the time lock for banks

  D. Being the first person to pick a double-acting lever tumbler lock

41. Which manufacturer is best known for its pushbutton combination locks?

   A. Kwikset Corporation

   B. Best Lock Corporation

   C. Better Lock and Interchangeable Core Company

   D. Simplex Access Controls

42. Which manufacturer is best known for its interchangeable core locks?

   A. Kwikset Corporation

   B. Best Lock Corporation

   C. Better Lock and Interchangeable Core Company

   D. Simplex Access Controls

43. Which manufacturer is best known for its residential key-in-knob locks?

   A. Kwikset Corporation

   B. Best Lock Corporation

   C. Better Lock and Interchangeable Core Company

   D. Simplex Access Controls

44. As a safety precaution, before servicing an ignition lock on an automobile, a locksmith should:

   A. Disconnect the negative battery cable from the battery

   B. Insert the proper key into the ignition lock

   C. Connect a ground cable to the automobile's battery

   D. Touch a plastic part of the vehicle to discharge static electricity

45. Ford began using wafer side bar locks in their automobiles in which year?

   A. 1984

   B. 1975

   C. 1981

   D. 1990

46. Which code series is commonly used by Ford glove compartment and trunk locks?

    A. "FMC" 2223–9999

    B. "FB" 0001–1863

    C. "GT" 1111–9999

    D. "FGT" 001–1888

47. The basic function of a CCTV's pan-and-tilt drive is to:

    A. Activate an alarm when it's touched by an intruder

    B. Point the camera

    C. Activate the videotape recorder

    D. Create lightness and darkness near the monitor as needed

48. General Motors ignition-lock codes can generally be found:

    A. On the ignition lock

    B. On the passenger-side door lock

    C. Below the Vehicle Identification Number (VIN) on the vehicle's engine

    D. Under the vehicle's brake pedal

49. How many styles of lock pawls does General Motors use in its various car lines?

    A. Three

    B. Over 20

    C. Two

    D. Five

50. Which code series is commonly used on Chrysler door and ignition locks?

    A. EP 1–3000

    B. CHR 1–5000

    C. CRY 1–4000

    D. GM 001–6000

51. Prior to 1993, there were 17 different resistor values for General Motors' VATS or PASSKey systems.

    A. True   B. False

52. Panic (life safety) exit devices usually have dogging,

    A. True   **B.** False

53. The Schlage J key blank enters the E, EF, F, and P keyways.

    A. True   **B.** False

54. The Abus Diskus padlock has a semi-round shackle.

    A. True   **B.** False

55. A safe's spline key holds the drive cam to the spindle.

    A. True   **B.** False

56. Schlage Lock Company uses a key cut depth increment of .0005 inch.

    A. True   **B.** False

57. The Schlage *D* Series lockset is designed primarily for light residential applications.

    A. True   **B.** False

58. Southern Steel is a major manufacturer of prison/detention locks and hardware.

    A. True   **B.** False

59. The rotating constant method of masterkeying requires change keys to have at least four bittings in the same position shared with the TMK.

    A. True   **B.** False

60. A typical disc-tumbler lock has at least nine different tumbler depths.

    A. True   **B.** False

Exam 3

1. City codes often dictate the height and style of fences.

   A. True   **B.** False

2. A hollow-core door is easy to break through.

   A. True   **B.** False

3. In general, peepholes should be installed on windowless exterior doors.

   A. True   **B.** False

4. A high-security strike box (or box strike) makes a door harder to kick in than a standard strike plate.

   A. True   **B.** False

5. A skeleton key can be used to open warded bit-key locks.

   A. True   **B.** False

6. If possible, house numbers should be visible from the street.

   A. True   **B.** False

7. A standard electromagnetic lock includes a rectangular electromagnet and a rectangular wood and glass strike plate.

   A. True   **B.** False

8. A *blank* is a key that fits two or more locks.

   A. True   **B.** False

9. One difference between a bit key and barrel key is the bit key has a hollow shank.

   A. True   **B.** False

10. A key-in-knob lock typically is used to secure windows.

    A. True   **B.** False

11. The Egyptians are credited with inventing the first lock to be based on the locking principle of today's pin-tumbler lock.

    A. True   **B.** False

12. A jimmy-proof deadlock typically is the most secure type of lock for sliding-glass doors.

    A. True   **B.** False

13. Lock picking is the most common way homes are burglarized.

    A. True   **B.** False

14. Lock impressioning is the most common way homes are burglarized.

    A. True   **B.** False

15. A long-reach tool and a wedge are commonly used to open locked automobiles.

    A. True   **B.** False

16. It is legal for locksmiths to duplicate a U.S. Post Office box key at the request of the box renter—if the box renter shows a current passport or driver's license.

    A. True   **B.** False

17. The Romans are credited with inventing the pin-tumbler lock.

    A. True   **B.** False

18. Five common keyway groove shapes are left angle, right angle, square, *V*, and round.

    A. True   **B.** False

19. To pick open a pin-tumbler cylinder, you usually need a pick and a torque wrench.

    A. True   **B.** False

20. A door reinforcer makes a lock harder to pick open.

    A. True   **B.** False

21. Common door-lock backsets include:

    A. 2½ inch and 2⅜ inch

    B. 3 inch and 2¾ inch

    C. ¾ inch and 2⅞ inch

    D. 2⅜ inch and 2¾ inch

22. How many sets of pin tumblers are in a typical pin-tumbler house door lock?

    A. Three or four

    B. Five or six

    C. Eleven or twelve

    D. One or eight

23. Which lock is unpickable?

    A. A Medeco biaxial deadbolt

    B. A Grade 2 Titan

    C. The Club steering wheel lock

    D. None of the above

24. Which are basic parts of a standard key-cutting machine?

    A. A pair of vises, a key stop, and a grinding stylus

    B. Two cutter wheels, a pair of vises, and a key shaper

    C. A pair of vises, a key stylus, and a cutter wheel

    D. A pair of styluses, a cutter wheel, and a key shaper

25. What are two critical dimensions for code-cutting cylinder keys?

    A. Spacing and depth

    B. Bow size and blade thickness

    C. Blade width and keyhole radius

    D. Shoulder width and bow size

26. Which manufacturer is best known for its low-cost residential key-in-knob locks?

    A. Kwikset Corporation

    B. Medeco Security Locks

    C. The Key-in-Knob Corporation

    D. ASSA

27. The most popular mechanical lock brands in the United States include:

    A. Yale, Master, Corby, and Gardall

    B. Yale, Kwikset, Master, and TuffLock

    C. Master, Weiser, Kwikset, and Schlage

    D. Master, Corby, Gardall, and TuffLock

28. A mechanical lock operated mainly by a pin-tumbler cylinder is commonly called:

    A. A disk-tumbler pinned lock

    B. A cylinder-pin lock

    C. A mechanical cylinder-pin lock

    D. A pin-tumbler cylinder lock

29. Burglars target garage doors because:

    A. People keep property that is easy to fence in garages.

    B. A garage attached to the house provides a discreet way to break into the house.

    C. A garage door with thin or loose panels can be accessed without opening the door.

    D. All of the above

30. Glass is a deterrent to burglars because

    A. It slows down a burglar.

    B. Broken shards of glass can injure a burglar.

    C. Shattering glass is noisy and attracts attention.

    D. All of the above

31. If you don't feel secure about glass in a window, you can increase security by:

   A. Replacing the glass with carbonated glass or antihammer plastic

   B. Replacing the glass with impact-resistant acrylic or polycarbonate, or with high-security glass

   C. Covering the glass with bullet-proof paint

   D. All of the above

32. Warded bit-key locks:

   A. Provide high security

   B. Provide little security

   C. Are hard to open without the right key

   D. Are usually the best choice for use on an exterior door

33. Which of the following key combinations provides the most security?

   A. 55555

   B. 33333

   C. 243535

   D. 35353

34. Which of the following key combinations provides the least security?

   A. 243535

   B. 1111

   C. 321231

   D. 22222

35. A blank is basically just:

   A. A change key with cuts on one side only

   B. An uncut or uncombinated key

   C. Any key with no words or numbers on the bow

   D. A master key with no words or numbers on the bow

36. You often can determine the number of pin stacks or tumblers in a cylinder by:

   A. Its key-blade length

   B. Its key-blade thickness

   C. The key-blank manufacturer's name on the bow

   D. The material of the key

37. Spool and mushroom pins:

   A. Make keys easier to duplicate

   B. Can hinder normal picking attempts

   C. Make a lock easier to pick

   D. Make keys harder to duplicate

38. As a general rule, the 10-cut wafer sidebar locks by General Motors have:

   A. A sum total of cut depths that must equal an even number

   B. Up to four of the same depth cut in the 7, 8, 9, and 10 spaces

   C. A maximum of five number-1 depths in a code combination

   D. At least one 4–1 or 1–4 adjacent cuts

39. When drilling open a standard pin-tumbler cylinder, position the drill bit:

   A. At the first letter of the cylinder

   B. At the shear line, in alignment with the top and bottom pins

   C. Directly below the bottom pins

   D. Directly above the top pins

40. When viewed from the exterior side, a door that opens inward and has hinges on the right side is:

   A. A left-hand door

   B. A right-hand door

   C. A left-hand reverse-bevel door

   D. A right-hand reverse-bevel door

41. A utility patent:

    A. Relates to a product's appearance, is granted for 14 years, and is renewable

    B. Relates to a product's function, is granted for 17 years, and is nonrenewable

    C. Relates to a product's appearance, is granted for 17 years, and is renewable

    D. Relates to a product's function, is granted for 35 years, and is nonrenewable

42. To earn a UL-437 rating, a sample lock must:

    A. Pass a performance test

    B. Use a patented key

    C. Use hardened-steel mounting screws, and mushroom and spool pins

    D. Pass an attack test using common hand and electric tools, such as drills, saw blades, puller mechanisms, and picking tools

43. Tumblers are:

    A. Small metal objects that protrude from a lock's cam to operate the bolt

    B. Fixed projections on a lock's case

    C. Small pins, usually made of metal, that move within a lock's case to prevent unauthorized keys from entering the keyhole

    D. Small objects, usually made of metal, that move within a lock cylinder in ways that obstruct a lock's operation until an authorized key or combination moves them into alignment

44. Electric switch locks:

    A. Are mechanical locks that have been modified to operate with battery power

    B. Complete and break an electric current when an authorized key is inserted and turned

    C. Are installed in metal doors to give electric shocks to intruders

    D. Are mechanical locks that have been modified to operate with alternating current (AC) electricity instead of with a key

45. A popular type of lock used on GM cars is:

    A. A Medeco pin tumbler

    B. An automotive bit key

    C. A sidebar wafer

    D. An automotive tubular key

46. When cutting a lever-tumbler key by hand, the first cut should be the:

    A. Lever cut

    B. Stop cut

    C. Throat cut

    D. Tip cut

47. How many possible key changes does a typical disk-tumbler lock have?

    A. 1,500

    B. 125

    C. A trillion

    D. 25

48. Which manufacturer is best known for its interchangeable core locks?

    A. Best Lock

    B. Kwikset Corporation

    C. ILCO Interchangeable Core Corporation

    D. Interchangeable Core Corporation

49. James Sargent is famous for:

    A. Inventing the Sargent key-in-knob lock

    B. Inventing the time lock for banks

    C. Inventing the double-acting lever-tumbler lock

    D. Being the first person to pick open a Medeco biaxial cylinder

50. Which are common parts of a combination padlock?

    A. Shackle, case, and bolt

    B. Spacer washer, top pins, and cylinder housing

    C. Back cover plate, case, and bottom pins

    D. Wheel-pack base plate, wheel pack spring, and top and bottom pins

51. General Motors' ignition lock codes generally can be found:

   A. On the ignition lock

   B. On the passenger-side door

   C. Below the Vehicle Identification Number (VIN) on the vehicle's engine

   D. Under the vehicle's brake pedal

52. Which code series is used commonly on Chrysler door and ignition locks?

   A. EP 1–3000

   B. CHR 1–5000

   C. CRY 1–4000

   D. GM 001–6000

53. How many styles of lock pawls does General Motors use in its various car lines?

   A. One

   B. Five

   C. More than five

   D. Three

54. The double-sided (or 10-cut) Ford key:

   A. Has five cuts on each side. One side operates the trunk and door, and the other side operates the ignition.

   B. Has five cuts on each side. Either side can operate all locks of a car.

   C. Has ten cuts on each side. One side operates the trunk and door, and the other side operates only the ignition.

   D. Has ten cuts on each side.

55. Usually the simplest way to change the combination of a double-bitted cam lock is to:

   A. Rearrange the positions of two or more tumblers

   B. Remove two tumblers and replace them with new tumblers

   C. Remove the tumbler assembly and replace it with a new one

   D. Connect a new tumbler assembly to the existing one

56. When shimming a pin-tumbler cylinder open:

   A. Use the key to insert the shim into the keyway

   B. Insert the shim into the keyway without the key

   C. Insert the shim along the left side of the cylinder housing

   D. Insert the shim between the plug and cylinder housing between the top and bottom pins

57. A lock is any:

   A. Barrier or closure that restricts entry

   B. Fastening device that allows a person to open and close a door, window, cabinet, drawer, or gate

   C. Device that incorporates a bolt, cam, shackle, or switch to secure an object—such as a door, drawer, or machine—to a closed, locked, on, or off position, and that provides a restricted means—such as a key or combination—of releasing the object from that position

   D. Device or object that restricts entry to a given premise

58. Which wheel in a safe lock is closest to the dial?

   A. Wheel 1

   B. Wheel 2

   C. Wheel 3

   D. Wheel 0

59. Which type of safe is best for protecting paper documents from fire?

   A. A paper safe

   B. A money safe

   C. A UL-listed record safe

   D. A patented burglary safe

60. Which type of cylinder is typically found on an interlocking deadbolt (or jimmy-proof dead-lock)?

   A. A bit-key cylinder

   B. A key-in-knob cylinder

   C. A rim cylinder

   D. A tubular deadbolt cylinder

Appendix F

REGISTERED SECURITY PROFESSIONAL EXAM

This test is based on the International Association of Home Safety and Security Professionals' Registered Security Professional Registration Program. If you earn a passing score, you should be able to pass other security certification and licensing examinations.

1. City codes often dictate the height and style of fences.

   A. True   **B.** False

2. A hollow-core door is easy to break through.

   A. True   **B.** False

3. In general, peepholes should be installed on windowless exterior doors.

   A. True   **B.** False

4. A high-security strike box (or box strike) makes a door harder to kick in than a standard strike plate.

   A. True   **B.** False

5. A skeleton key can be used to open warded bit-key locks.

   A. True   **B.** False

6. If possible, house numbers should be visible from the street.

   A. True   **B.** False

7. A standard electromagnetic lock includes a rectangular electromagnet and a rectangular wood and glass strike plate.

   A. True   **B.** False

8. A *blank* is a key that fits two or more locks.

   A. True   **B.** False

9. One difference between a bit key and a barrel key is the bit key has a hollow shank.

   A. True   **B.** False

10. A key-in-knob lock typically is used to secure windows.

    A. True   **B.** False

11. The Egyptians are credited with inventing the first lock to be based on the locking principle of today's pin-tumbler lock.

    A. True   **B.** False

12. A jimmy-proof deadlock typically is the most secure type of lock for sliding-glass doors.

    A. True   **B.** False

13. Lock picking is the most common way homes are burglarized.

    A. True   **B.** False

14. Lock impressioning is the most common way homes are burglarized.

    A. True   **B.** False

15. A long-reach tool and a wedge are commonly used to open locked automobiles.

    A. True   **B.** False

16. It's legal for locksmiths to duplicate a U.S. Post Office box key at the request of the box renter— if the box renter shows a current passport or driver's license.

    A. True   **B.** False

17. The Romans are credited with inventing the pin-tumbler lock.

    A. True   **B.** False

18. Five common keyway groove shapes are left angle, right angle, square, *V*, and round.

    A. True   **B.** False

19. To pick open a pin-tumbler cylinder, you usually need a pick and a torque wrench.

    A. True   **B.** False

20. A door reinforcer makes a lock harder to pick open.

    A. True   **B.** False

21. Common door-lock backsets include:

    A. 2½ inch and 3 inch

    B. 3 inch and 2¾ inch

    C. 1¾ inch and 2½ inch

    D. 2 inch and 2¾ inch

22. How many sets of pin tumblers are in a typical pin-tumbler house door lock?

    A. Three or four

    B. Five or six

    C. Eleven or twelve

    D. Seven or eight

23. Which lock is unpickable?

    A. A Medeco biaxial deadbolt

    B. A Grade 2 Titan

    C. The Club steering wheel lock

    D. None of the above

24. Which are basic parts of a standard key-cutting machine?

    A. A pair of vises, a key stop, and a grinding stylus

    B. Two cutter wheels, a pair of vises, and a key shaper

    C. A pair of vises, a key stylus, and a cutter wheel

    D. A pair of styluses, a cutter wheel, and a key shaper

25. What are two critical dimensions for code-cutting cylinder keys?

    A. Spacing and depth

    B. Bow size and blade thickness

    C. Blade width and keyhole radius

    D. Shoulder width and bow size

26. Which manufacturer is best known for its low-cost residential key-in-knob locks?

    A. Kwikset Corporation

    B. Medeco Security Locks

    C. The Key-in-Knob Corporation

    D. ASSA

27. The most popular mechanical lock brands in the United States include:

    A. Yale, Master, Corby, and Gardall

    B. Yale, Kwikset, Master, and TuffLock

    C. Master, Weiser, Kwikset, and Schlage

    D. Master, Corby, Gardall, and TuffLock

28. A mechanical lock that is operated mainly by a pin-tumbler cylinder is commonly called a:

    A. Disk-tumbler pinned lock

    B. Cylinder-pin lock

    C. Mechanical cylinder-pin lock

    D. Pin-tumbler cylinder lock

29. Burglars target garage doors because:

    A. People keep property that is easy to fence in garages.

    B. A garage that's attached to the house provides a discreet way to break into the house.

    C. A garage door with thin or loose panels can be accessed without opening the door.

    D. All of the above

30. Glass is a deterrent to burglars because:

    A. It slows down a burglar.

    B. Broken shards of glass can injure a burglar.

    C. Shattering glass is noisy and attracts attention.

    D. All of the above

31. If you don't feel secure about glass in a window, you can increase security by:

    A. Replacing the glass with carbonated glass or antihammer plastic

    B. Replacing the glass with impact-resistant acrylic or polycarbonate or high-security glass

    C. Covering the glass with bullet-proof paint

    D. All of the above

32. Warded bit-key locks:

    A. Provide high security

    B. Provide little security

    C. Are hard to open without the right key

    D. Are usually the best choice for use on an exterior door

33. Which of the following key combinations provides the most security?

    A. 55555

    B. 33333

    C. 243535

    D. 35353

34. Which of the following key combinations provides the least security?

    A. 243535

    B. 1111

    C. 321231

    D. 22222

35. A blank is basically just:

    A. A change key with cuts on one side only

    B. An uncut or uncombinated key

    C. Any key with no words or numbers on the bow

    D. A master key with no words or numbers on the bow

36. You often can determine the number of pin stacks or tumblers in a cylinder by:

    A. Its key-blade length

    B. Its key-blade thickness

    C. The key-blank manufacturer's name on the bow

    D. The material of the key

37. Spool and mushroom pins:

    A. Make keys easier to duplicate

    B. Can hinder normal picking attempts

    C. Make a lock easier to pick

    D. Make keys harder to duplicate

38. As a general rule, General Motors' 10-cut wafer sidebar locks have:

    A. A sum total of cut depths that must equal an even number

    B. Up to four of the same depth cut in the 7, 8, 9, and 10 spaces

    C. A maximum of five number-1 depths in a code combination

    D. At least one 4–1 or 1–4 adjacent cuts

39. When drilling open a standard pin-tumbler cylinder, position the drill bit:

    A. At the first letter of the cylinder

    B. At the shear line in alignment with the top and bottom pins

    C. Directly below the bottom pins

    D. Directly above the top pins

40. When viewed from the exterior side, a door that opens inward and has hinges on the right side is a:

    A. Left-hand door

    B. Right-hand door

    C. Left-hand reverse-bevel door

    D. Right-hand reverse-bevel door

41. A utility patent:

    A. Relates to a product's appearance, is granted for 14 years, and is renewable

    B. Relates to a product's function, is granted for 17 years, and is nonrenewable

    C. Relates to a product's appearance, is granted for 17 years, and is renewable

    D. Relates to a product's function, is granted for 35 years, and is nonrenewable

42. To earn a UL-437 rating, a sample lock must:

    A. Pass a performance test

    B. Use a patented key

    C. Use hardened-steel mounting screws, and mushroom and spool pins

    D. Pass an attack test using common hand and electric tools such as drills, saw blades, puller mechanisms, and picking tools

43. Tumblers are:

    A. Small metal objects that protrude from a lock's cam to operate the bolt

    B. Fixed projections on a lock's case

    C. Small pins, usually made of metal, that move within a lock's case to prevent unauthorized keys from entering the keyhole

    D. Small objects, usually made of metal, that move within a lock cylinder in ways that obstruct a lock's operation until an authorized key or combination moves them into alignment

44. Electric switch locks:

    A. Are mechanical locks that have been modified to operate with battery power

    B. Complete and break an electric current when an authorized key is inserted and turned

    C. Are installed in metal doors to give electric shocks to intruders

    D. Are mechanical locks that have been modified to operate with alternating-current (AC) electricity instead of with a key

45. A popular type of lock used on GM cars is:

    A. A Medeco pin tumbler

    B. An automotive bit key

    C. A sidebar wafer

    D. An automotive tubular key

46. When cutting a lever-tumbler key by hand, the first cut should be the:

   A. Lever cut

   B. Stop cut

   C. Throat cut

   D. Tip cut

47. How many possible key changes does a typical disk-tumbler lock have?

   A. 1,500

   B. 125

   C. A trillion

   D. 25

48. Which manufacturer is best known for its interchangeable core locks?

   A. Best Lock

   B. Kwikset Corporation

   C. ILCO Interchangeable Core Corporation

   D. Interchangeable Core Corporation

49. James Sargent is famous for:

   A. Inventing the Sargent key-in-knob lock

   B. Inventing the time lock for banks

   C. Inventing the double-acting lever-tumbler lock

   D. Being the first person to pick open a Medeco biaxial cylinder

50. Which are common parts of a combination padlock?

   A. Shackle, case, and bolt

   B. Spacer washer, top pins, and cylinder housing

   C. Back cover plate, case, and bottom pins

   D. Wheel-pack base plate, wheel pack spring, and top and bottom pins

51. General Motors' ignition lock codes generally can be found:

   A. On the ignition lock

   B. On the passenger-side door

   C. Below the Vehicle Identification Number (VIN) on the vehicle's engine

   D. Under the vehicle's brake pedal

52. Which code series is used commonly on Chrysler door and ignition locks?

   A. EP 1–3000

   B. CHR 1–5000

   C. CRY 1–4000

   D. GM 001–6000

53. How many styles of lock pawls does General Motors use in its various car lines?

   A. One

   B. Five

   C. Over five

   D. Three

54. The double-sided (or 10-cut) Ford key:

   A. Has five cuts on each side. One side operates the trunk and door, and the other side operates the ignition.

   B. Has five cuts on each side. Either side can operate all locks of a car.

   C. Has ten cuts on each side. One side operates the trunk and door, and the other side operates only the ignition.

   D. Has ten cuts on each side.

55. Usually, the simplest way to change the combination of a double-bitted cam lock is to:

   A. Rearrange the positions of two or more tumblers

   B. Remove two tumblers and replace them with new tumblers

   C. Remove the tumbler assembly and replace it with a new one

   D. Connect a new tumbler assembly to the existing one

56. When shimming a pin-tumbler cylinder open:

    A. Use the key to insert the shim into the keyway.

    B. Insert the shim into the keyway without the key.

    C. Insert the shim along the left side of the cylinder housing.

    D. Insert the shim between the plug and cylinder housing between the top and bottom pins.

57. A lock is any:

    A. Barrier or closure that restricts entry

    B. Fastening device that allows a person to open and close a door, window, cabinet, drawer, or gate

    C. Device that incorporates a bolt, cam, shackle, or switch to secure an object—such as a door, drawer, or machine—to a closed, locked, on, or off position and that provides a restricted means—such as a key or combination—of releasing the object from that position

    D. Device or object that restricts entry to a given premise

58. Which wheel in a safe lock is closest to the dial?

    A. Wheel 1

    B. Wheel 2

    C. Wheels 3

    D. Wheel 0

59. Which type of safe is best for protecting paper documents from fire?

    A. A paper safe

    B. A money safe

    C. A UL-listed record safe

    D. A patented burglary safe

60. Which type of cylinder is typically found on an interlocking deadbolt (or jimmy-proof deadlock)?

    A. A bit-key cylinder

    B. A key-in-knob cylinder

    C. A rim cylinder

    D. A tubular deadbolt cylinder

Appendix G
ANSWERS TO EXAMS

Appendix B

1. B
2. A
3. B
4. D
5. A
6. A
7. B
8. B
9. A
10. C
11. C
12. B
13. B
14. C
15. E
16. C
17. B
18. A
19. C
20. B
21. C
22. C
23. D
24. C
25. C
26. C
27. D
28. A
29. A
30. A
31. C
32. C
33. A
34. D
35. A

Appendix C

1. B
2. A
3. B
4. B
5. A
6. B
7. B
8. B
9. A
10. B
11. A
12. B
13. B
14. A
15. B
16. B
17. B
18. B
19. A
20. B
21. B
22. B
23. A
24. B
25. A
26. A
27. B
28. B
29. B

30.	B		**64.**	B
31.	A		**65.**	A
32.	B		**66.**	A
33.	A		**67.**	A
34.	A		**68.**	A
35.	B		**69.**	B
36.	B		**70.**	B
37.	B		**71.**	B
38.	A		**72.**	A
39.	B		**73.**	B
40.	B		**74.**	B
41.	A		**75.**	A
42.	A		**76.**	B
43.	B		**77.**	B
44.	B		**78.**	B
45.	A		**79.**	A
46.	A		**80.**	B
47.	B		**81.**	A
48.	A		**82.**	A
49.	B		**83.**	B
50.	B		**84.**	B
51.	B		**85.**	B
52.	A		**86.**	A
53.	A		**87.**	A
54.	B		**88.**	B
55.	A		**89.**	B
56.	B		**90.**	A
57.	B		**91.**	A
58.	B		**92.**	B
59.	A		**93.**	A
60.	B		**94.**	B
61.	A		**95.**	B
62.	A		**96.**	B
63.	A		**97.**	B

98.	A			28.	A
99.	B			29.	B
100.	B			30.	C

Appendix D, Exam 1

Unit One

1.	B
2.	A
3.	A
4.	B
5.	B
6.	B
7.	D
8.	D
9.	B
10.	D
11.	B
12.	C
13.	D
14.	C
15.	B
16.	A
17.	B
18.	B
19.	A
20.	B
21.	B
22.	C
23.	C
24.	A
25.	B
26.	A
27.	D

Unit Two

31.	D
32.	B
33.	B
34.	A
35.	B
36.	D
37.	A
38.	E
39.	A
40.	A
41.	B
42.	A
43.	C
44.	A
45.	B
46.	B
47.	C
48.	B
49.	C
50.	B
51.	B
52.	C
53.	A
54.	A
55.	D
56.	A
57.	A
58.	A
59.	B

60.	D		**Unit Four**	
61.	A		91.	A
62.	A		92.	B
63.	B		93.	B
64.	A		94.	A
65.	A		95.	A
66.	D		96.	A
67.	D		97.	D
68.	D		98.	A
69.	A		99.	C
70.	D		100.	D
71.	A		101.	B
72.	A		102.	A
73.	D		103.	B
74.	D		104.	A
75.	D		105.	B
76.	B		106.	B
77.	B		107.	A
78.	B		108.	D
79.	B		109.	A
80.	A		110.	C
			111.	A
Unit Three			112.	A
81.	B		113.	C and D
82.	B		114.	A
83.	A		115.	C
84.	B		116.	B
85.	A		117.	C
86.	D		118.	B
87.	C		119.	C
88.	C		120.	A
89.	B		121.	C
90.	C		122.	B

123.	A
124.	B
125.	A
126.	B
127.	B
128.	A
129.	A
130.	B

Unit Five

131.	A
132.	A
133.	B
134.	B
135.	A
136.	D
137.	C
138.	E
139.	C
140.	C
141.	B
142.	B
143.	B
144.	A
145.	C
146.	C
147.	A
148.	D
149.	B
150.	B
151.	B
152.	A
153.	D
154.	D

155.	D
156.	C
157.	C
158.	B
159.	D
160.	D

Unit Six

161.	A
162.	B
163.	B
164.	D
165.	B
166.	B and C
167.	B
168.	P–C–A
169.	B
170.	D
171.	D
172.	B
173.	B
174.	B
175.	A
176.	B
177.	A
178.	D
179.	D
180.	D
181.	A
182.	A
183.	A
184.	B
185.	E
186.	D

187.	E		**219.**	D
188.	E		**220.**	A
189.	B		**221.**	E
190.	E		**222.**	A
191.	A		**223.**	A
192.	B		**224.**	B
193.	A		**225.**	B
194.	A		**226.**	A
195.	B		**227.**	D
196.	C		**228.**	A
197.	D		**229.**	C
198.	A		**230.**	B
199.	B		**231.**	B
200.	B		**232.**	D
			233.	A
			234.	D

Unit Seven

201.	B		**235.**	A
202.	B		**236.**	B
203.	B		**237.**	B
204.	B		**238.**	A
205.	A		**239.**	C
206.	A		**240.**	B
207.	C			
208.	A			

Unit Eight

209.	A		**241.**	D
210.	D		**242.**	B
211.	B		**243.**	A
212.	A		**244.**	B
213.	C		**245.**	B
214.	D		**246.**	B
215.	B		**247.**	B
216.	A		**248.**	B
217.	C		**249.**	B
218.	B		**250.**	D

Unit Nine

251.	B
252.	B
253.	A
254.	D
255.	B
256.	B
257.	A
258.	C
259.	B
260.	E

Unit Ten

261.	B
262.	A
263.	A
264.	A
265.	A
266.	B
267.	A
268.	D
269.	B
270.	B

Unit Eleven

271.	A
272.	B
273.	C
274.	B
275.	A
276.	B
277.	D
278.	C
279.	B
280.	D

Unit Twelve

281.	B
282.	B
283.	D
284.	A
285.	D
286.	D
287.	D
288.	C
289.	D
290.	B
291.	A
292.	A and B
293.	B
294.	A
295.	D
296.	B
297.	C
298.	A
299.	A
300.	A
301.	D
302.	B
303.	D
304.	B
305.	A
306.	B
307.	D
308.	D
309.	A
310.	B

Appendix E, Exam I

The multiple choice questions count as two points, and the True/False answers count as one point. To pass the test, 70 points are required.

1.	A
2.	A
3.	B
4.	B
5.	A
6.	B
7.	B
8.	B
9.	A
10	A
11.	A
12.	B
13.	A
14.	B
15.	A
16.	B
17.	A
18.	A
19.	A
20.	B
21.	A
22.	B
23.	D
24.	C
25.	A
26.	A
27.	C
28.	D
29.	C
30.	D
31.	B
32.	C
33.	C
34.	B
35.	B
36.	A
37.	B
38.	A
39.	B
40.	A
41.	B
42.	A
43.	D
44.	B
45.	C
46.	C
47.	B
48.	A
49.	B
50.	A
51.	A
52.	B
53.	C
54.	C
55.	C
56.	D
57.	C
58.	C
59.	B
60.	A

Appendix E, Exam 2

When scoring the test, give yourself two points for each correctly answered multiple choice question and one point for each correctly answered True/False question. In each case where you gave no answer to a question, and in each case where you gave two or more answers to a question, give yourself no credit.

Scoring Key: 75 percent is minimum for passing, 85 percent is very good, and 95 percent or better is excellent.

1.	A
2.	B
3.	A
4.	A
5.	D
6.	D
7.	B
8.	A
9.	A
10.	A
11.	B
12.	C
13.	D
14.	B
15.	D
16.	B
17.	A
18.	A
19.	A
20.	D
21.	A
22.	B
23.	B
24.	A
25.	D

26.	C
27.	C
28.	A
29.	C
30.	B
31.	A
32.	A
33.	D
34.	C
35.	B
36.	C
37.	B
38.	C
39.	C
40.	C
41.	D
42.	B
43.	A
44.	A
45.	A
46.	B
47.	B
48.	A
49.	B
50.	A
51.	B
52.	A
53.	A
54.	A
55.	A
56.	B
57.	B
58.	A
59.	B
60.	B

Appendix E, Exam 3

When scoring the test, give yourself two points for each correctly answered multiple choice question and one point for each correctly answered True/False question. In each case where you gave no answer to a question, and in each case where you gave two or more answers to a question, give yourself no credit.

Scoring Key: 75 percent is minimum for passing, 85 percent is very good, and 95 percent or better is excellent.

1.	B
2.	A
3.	A
4.	A
5.	A
6.	A
7.	B
8.	B
9.	B
10.	B
11.	A
12.	B
13.	B
14.	B
15.	A
16.	B
17.	B
18.	A
19.	A
20.	B
21.	A
22.	B
23.	D
24.	C
25.	A

26.	A
27.	C
28.	D
29.	D
30.	D
31.	D
32.	B
33.	C
34.	B
35.	B
36.	A
37.	B
38.	C
39.	B
40.	B
41.	B
42.	D
43.	D
44.	B
45.	C
46.	C
47.	B
48.	A
49.	B
50.	A
51.	A
52.	B
53.	C
54.	B
55.	C
56.	D
57.	C
58.	A
59.	C
60.	C

Appendix F

When scoring the test, give yourself two points for each correctly answered multiple choice question and one point for each correctly answered True/False question. In each case where you gave no answer to a question, and in each case where you gave two or more answers to a question, give yourself no credit.

Scoring Key: 75 percent is minimum for passing, 85 percent is very good, and 95 percent or better is excellent.

1.	A
2.	A
3.	A
4.	A
5.	A
6.	A
7.	B
8.	B
9.	B
10.	B
11.	A
12.	B
13.	B
14.	B
15.	A
16.	B
17.	B
18.	A
19.	A
20.	B
21.	D
22.	B
23.	D
24.	C
25.	A
26.	A
27.	C
28.	D
29.	D
30.	D
31.	B
32.	B
33.	C
34.	B
35.	B
36.	A
37.	B
38.	D
39.	B
40.	B
41.	B
42.	D
43.	D
44.	B
45.	C
46.	C
47.	B
48.	A
49.	B
50.	A
51.	A
52.	B
53.	C
54.	D
55.	C
56.	B
57.	C
58.	A
59.	C
60.	C

Appendix H
GLOSSARY

Alternating current. Electricity that runs through electrical outlets in a building.

Angularly bitted key. A key with cuts that angle perpendicularly from the blade. Such a key is often used to operate high-security locks, such as Medeco locks.

Antilift plate. A metal plate installed at the top of sliding glass doors to hinder attempts to lift the door out of the frame.

Backset. The distance between the center of a cross-bore and the bolt edge of a door or drawer.

Bit key. A key used for operating bit-key locks (sometimes called a *skeleton key*).

Blade. The side of a key that is meant to cut or mill.

Bow. (Rhymes with "toe.") The head of a key. The part a person grips when using a key.

Bump key. A key cut to let the user copy the action of a pick gun.

Burglary safe. A safe designed primarily to safeguard its contents from burglary. Also called a "money" safe.

Casement window. A window hinged on one side that swings outward like a door.

Class A fire extinguisher. Used for wood, paper, plastic, and clothing fires.

Class B fire extinguisher. Used for grease, gasoline, petroleum oil, and other flammable liquids fires.

Class C fire extinguisher. Used for electrical equipment and wiring fires.

Corrugated key. A key with corrugations, or ripples, along the length of its blades.

Cylinder. The part of a lock that consists of a plug, tumblers, and springs.

Cylinder lock. A lock that uses a cylinder key.

Deadbolt. A bolt that requires a deliberate action to extend or retract, and that cannot be moved to the unlocked position simply by applying end pressure to the bolt.

Deadbolt lock. A lock that contains a deadbolt.

Depository safe. A safe with a slot in it, so cashiers can insert money into it. The slot prevents people from taking the money out again without using a key or combination.

Detection devices. The eyes and ears of an alarm system. Detection devices sense the presence of an intruder and relay the information to the control panel. They are sometimes called "sensors."

Dimple key. A key that has cuts drilled or milled into its blade surface; the cuts normally don't change the blade's silhouette. Such a key is often used to operate high-security locks, such as DOM locks.

Direct current. Electricity that comes from batteries.

Door reinforcer. A metal piece that wraps around a door edge to strengthen the door against kick-ins.

Double-hung window. A window that consists of two square or rectangular sashes that slide up and down.

Electric strike. A strike that locks and unlocks by electric current.

Electromagnetic lock. A lock that relies on electricity and magnetism. It usually consists of a rectangular electromagnet and a rectangular, ferrous metal strike plate.

Fire safe. A safe designed to protect its contents from fire. Also called a "record" safe.

Flat steel key. A key that is flat on both sides.

Floor safe. A safe designed to sit on top of a floor. Some locksmiths refer to an in-floor safe as a floor safe.

Glazing. Any transparent or translucent material—usually some kind of glass or plastic—used on windows or doors to let in light.

Impressioning. A technique for opening a lock by binding a key blank in the keyway to make tumbler impressions on the blank, and then filing the blank until a working key is made.

In-floor safe. A safe designed to be installed below the surface of a floor.

Jamb (of a door). The vertical part of a door frame.

Louvered window. A window made of a ladder-like configuration of narrow, overlapping slats of glass that can easily be pulled out of the thin metal channels. Also called a "jalousie" window.

Key control. Any method of preventing unauthorized duplication of a key.

Key-in-knob lockset. A lockset that uses one or more knobs.

Key-in-lever lockset. A lockset that uses one or more lever handles.

Lock. A device that incorporates a bolt, shackle, or switch to secure an object—such as a door, drawer, or machine—to a closed, opened, locked, or an on or off position, and provides a restricted means of releasing the object from that position.

Lock pick. A tool designed to enter a keyway and move tumblers to the unlocked position.

Locksmith. A person whose job is to install and service locks.

Mortise cylinder. A threaded cylinder used in mortise locks.

Pick gun. A tool designed to quickly slap tumblers to the unlocked position.

Plug. The part of a lock that consists of a keyway and holds tumblers.

Slim jim. A flat piece of metal, usually one to two inches wide, used for opening locked automobiles by being inserted in a door cavity.

Static electricity. Electricity that happens in one place instead of running through circuits (for example, the shock experienced when a shoe is rubbed across carpet; lightning).

Tension wrench. A tool used to apply turning pressure to a lock, while using a pick to open it. Also called a "torque" wrench.

Tubular key. A key that has a tubular blade with cuts, or depressions, milled in a circle around the end of the blade. Sometimes called an "Ace key."

Tumbler. A movable obstruction that comes into contact with the cuts and/or millings of keys in such a way that the lock can be unlocked or locked.

Ventilating wood window lock. An L-shaped metal-bolt assembly and a small metal base that lets someone inside raise a double-hung window a few inches, and then set the bolt, which prevents anyone from outside from raising the window higher.

Appendix I
BIBLIOGRAPHY

Capel, Vivian. *Security Systems and Intruder Alarms*. Portsmouth, NH: Heinemann-Newnes, 1989.

Craighead, Geoff. *High-Rise Security and Fire Life Safety*, 2nd ed. Burlington, MA: Butterworth-Heinemann, 2003.

Crowe, Timothy D. *Crime Prevention Through Environmental Design*, 2nd ed. Burlington, MA: Butterworth-Heinemann, 2000.

Cumming, Neil. *Security*, 2nd ed. Burlington, MA: Butterworth-Heinemann, 1992.

Dalton, Dennis R. *Rethinking Corporate Security in the Post 9/11 Era*. Burlington, MA: Butterworth-Heinemann, 2003.

Field, Frank, Dr., and John Morse. *Dr. Frank Field's Get Out Alive*. New York: Random House, 1992.

Fischer, Robert J., and Gion Green. *Introduction to Security*, 7th ed. Burlington, MA: Butterworth-Heinemann, 2004.

Garcia, Mary Lynn. *The Design and Evaluation of Physical Protection Systems*. Burlington, MA: Butterworth-Heinemann, 2001.

Honey, Gerard. *Intruder Alarms*, 2nd ed. Oxford, England: Elsevier Science, 2003.

Muuss, James P., CPP, and David Rabern, CPP. *The Complete Guide for CPP Examination Preparation*. New York: Auerbach Publications, 2006.

National Crime Prevention Institute. *Understanding Crime Prevention*. Burlington, MA: Butterworth-Heinemann, 2001.

Phillips, Bill. *Master Locksmithing*. New York: McGraw-Hill, 2008.

Phillips, Bill. *The Complete Book of Home, Site, and Office Security*. New York: McGraw-Hill, 2008.

Robinson, Robert L. *Complete Course in Professional Locksmithing*. Chicago: Nelson-Hall, 1973.

Sennnewald, Charles A. *Effective Security Management*, 4th ed. Burlington, MA: Butterworth-Heinemann, 2003.

Thomas, Gerry S. *Business Handbook on Terrorism and Security Survival*. Alexandria, VA: Vance Brook Publishing, 1995.

Tobias, Marc Weber, J.D. *Locks, Safes, and Security*, 2nd ed. Springfield, IL: Charles C. Thomas Publisher, Ltd., 2000.

Traister, John E. *Security/Fire-Alarm Systems*, 2nd ed. New York: McGraw-Hill, 1996.